Climate Change in an Aging Society

Climate Change in an Aging Society is the first book fully devoted to the impact of climate change on those who are old today—and those who will be old in decades to come. In doing so, Moody focuses on issues of critical importance: aging in place; health and age in a warming world; responsibility for the climate crisis; options for climate-conscious consumers; planning for investment for a green retirement; and opportunities for political action.

The number of Americans aged over 65 is projected to rise from 17% to 21%. By 2060 nearly one in four Americans will be 65 or older. By 2050, however, average temperatures in the USA could rise by as much as 3°C, and extreme weather events are likely to become more frequent and severe. Despite these alarming projections and the likelihood that climate change will cause serious health issues among the elderly, little attention has been devoted to the impact of climate change on this demographic. Employing a life-course perspective and a cross-generational approach, Moody assesses the impact of climate change on those who are old today and those who will be old in years to come. Challenging both climate complacency and climate defeatism, the book adopts as its clarion call, HERE NOW YOU HOPE.

Written in an engaging personal style with highlighting case studies of influential "eco-elders," this urgent book will be of great interest to students and scholars with interests in climate change, gerontology, and environmental and social policy.

Harry R. Moody is Visiting Faculty in the Creative Longevity and Wisdom Program of Fielding Graduate University, USA. He is former Vice President and Director of Academic Affairs at AARP in Washington and is currently the editor of the popular *Human Values in Aging* monthly newsletter. In 2011, he received the Lifetime Achievement Award from the American Society on Aging, and in 2008 he was named by *Utne Reader* Magazine as one of "50 Visionaries Who Are Changing Your World." He is the author of *Ethics in an Aging Society* (1992), *Abundance of Life: Human Development Policies for an Aging Society* (1988), and over 100 scholarly articles. He is also the co-author of *Aging: Concepts and Controversies* (10th edition, 2020), *Gerontology: The Basics* (Routledge, 2018), *Dignity and Old Age* (Routledge, 1998), and *The Five Stages of the Soul: Charting the Spiritual Passages That Shape Our Lives* (1997), and the editor of *Religion, Spirituality, and Aging: A Social Work Perspective* (Routledge, 2005).

Aging and Society
Edited by Carroll L. Estes and
Assistant Editor Nicholas DiCarlo

This pioneering series of books creatively synthesizes and advances key, intersectional topics in gerontology and aging studies. Drawing from changing and emerging issues in gerontology, influential scholars combine research into human development and the life course; the roles of power, policy, and partisanship; race and ethnicity; inequality; gender and sexuality; and cultural studies to create a multi-dimensional and essential picture of modern aging.

When Strangers Become Family
The Role of Civil Society in Addressing the Needs of Aging Populations (2021)
Ronald Angel and Verónica Montes-de-Oca Zavala

Safeguarding Social Security for Future Generations
Leaving a Legacy in an Aging Society
W. Andrew Achenbaum

Long Lives are for the Rich Aging, the Life Course, and Social Justice
Jan Baars

Ageing in Place in Urban Environments Critical Perspectives
Tine Buffel and Chris Phillipson

Why Place Matters Place and Place Attachment for Older Adults
Joyce Weil

Care Justice Reframing Public Policy, Elevating Care Work
Nancy Hooyman

Climate Change in an Aging Society
Harry R. Moody

For more information about this series, please visit: www.routledge.com/Aging-and-Society/book-series/AGINGSOC

Climate Change in an Aging Society

Harry R. Moody

Routledge
Taylor & Francis Group

NEW YORK AND LONDON

Designed cover image: Getty Images (RapidEye)

First published 2025
by Routledge
605 Third Avenue, New York, NY 10158

and by Routledge
4 Park Square, Milton Park, Abingdon, Oxon, OX14 4RN

Routledge is an imprint of the Taylor & Francis Group, an informa business

ISBN: 978-1-032-38637-9 (hbk)
ISBN: 978-1-032-38636-2 (pbk)
ISBN: 978-1-003-34599-2 (ebk)

DOI: 10.4324/9781003345992

Typeset in Sabon
by Apex CoVantage, LLC

Contents

Boxes

Figures

Preface

Hope and Fear in the Face of Climate Change: That's the book you hold in your hands.

This is a book about climate change for people who are already old, like the author. It's also for people who hope to grow old, like most readers of this book. It offers a simple message: HERE NOW YOU HOPE—just four words. Climate change is already here and now. If climate change is not yet having an impact on when and where you live at this moment, it will have an impact before long. It's here and now, and it affects you and those you care about.

In short, climate change is not about a distant future but about what will happen soon enough, for you and the people you care about. If you don't want to think about what will happen, just stop right here. Don't read further. Also, don't bother to read that biopsy report from your doctor. Don't look at your tax notices or credit cards. Don't pay attention to anything that might disturb you.

Anyone who gets the message HERE NOW and YOU might become depressed, maybe even want to go on a "news fast" and then say "I don't follow the news." Some will run away from climate anxiety. But they won't run far enough. They need to read the fourth word—HOPE. If you have already heard the bad news about climate change, you may not yet have heard the word HOPE.

The meaning of hope is not the same as optimism, and it's not a form of magical thinking. Magical thinking is what we see with Mr. Micawber in Charles Dickens's novel *David Copperfield*, the man who always says, "Something will turn up." It's also the strategy for Scarlett O'Hara (*Gone with the Wind*), the woman who says, "After all, tomorrow is another day!" Magical thinking is a survival tool but a path to disaster. It's not the tool you need.

Most live our lives by blind habit, by custom, and by complacency. Thoreau said, "Most people live lives of quiet desperation." For many climate change is just one more source of desperation. We need something more, and that is HOPE. The word HOPE means something different when it comes to climate action. It was stated best by environmentalist David Orr: "Hope is a verb with its sleeves rolled up." Hope is not optimism and not evasion. It is the basis for action. That's what this book is about: to see how climate change affects us, as we grow older. And then to see what we can do about it, how to roll up your sleeves and help yourself and those around you. In short, it's not too late, as Rebecca Solnit insists.[1]

Can hope be an illusion, or can it be false hope? Yes, it can be an illusion, as we know too well from people who thought they were saving for retirement by trusting Bernard Madoff with their money. When it comes to growing older, yes, hope can be an illusion if you follow the advice of anti-aging doctors who prescribe medications promising to cure old age. I've personally known several people who died after taking medications promising false hope.

When it comes to money or health, we need answers that take account of risks and also point us to actions we can take for a better future. So, yes, fear has its proper place in commanding our attention, in waking us from complacency.

Greta Thunberg, born in 2003, is not an elder. Thunberg is the Swedish climate activist who founded Fridays for Future, also called School Strike for Climate. Thunberg wants us to be afraid, and she says it bluntly:

> I don't want your hope. I don't want you to be hopeful. I want you to panic. I want you to feel the fear I feel every day. And then I want you to act. I want you to act as you would in a crisis. I want you to act as if our house is on fire. Because it is.[2]

Yes, our house is on fire, and we need to act. Greta Thunberg is right, even if she doesn't tell us the whole story. The real story means we act only if we have hope and if hope turns into "rolling up our sleeves" to do what is needed. Like fear, the right hope is also what we need.

In this book, we look at the tangible threats, the reasons for fear: wildfire, flood, heat wave, and drought. We also look at who caused this mess, at who is to blame, ranging from those who are rich to older people like me, including even readers of this book. The point of blame is not guilt. It's to get rid of excuses and get ready for action, for rolling up our sleeves. We do it by becoming climate-conscious consumers, investing in a green retirement, and joining with others in collective action to help ourselves and to save the planet. When faced with the question, "What can one person do?" we have the answer: Don't just act as one person.

I'm 79 years old, and I've been working in the "aging business" since 1971: more than half a century, most of my life: in Senior Centers, higher education (Hunter College), AARP, and at Fielding Graduate University. As we like to say in gerontology, I have been "aging in place." Eventually, by growing older, I have become a specimen for what I was studying in gerontology. Trained first as a philosopher who later became a gerontologist, I see that I came to have one prime purpose: positive aging.

Now I've turned my attention to climate and for good reason. Unless we can fix the platform where we're living—planet Earth—we have no hope for a good old age. For our successors, things could be worse. All generations are at risk, those old today, like me, and those who hope to grow old, like my children and grandchildren.

Those of us with normal vision see with two eyes. We will act only on climate change when we see with the two eyes of the heart: that is, when we see with both fear and hope. No book can change the world. Complacency tempts us to turn the page and go back to sleep, back to habits of everyday life, back to quiet desperation, or back to the false hope that something will just turn up. We can do better than this, and that's what you'll read in the pages ahead.

Notes

1 Rebecca Solnit and Thelma Lutunatabua, *Not Too Late: Changing the Climate Story from Despair to Possibility* (Haymarket Books, 2023).
2 " 'Our House Is on Fire': Greta Thunberg, 16, Urges Leaders to Act on Climate" *The Guardian* (Jan. 25, 2019) at www.theguardian.com/environment/2019/jan/25/our-house-is-on-fire-greta-thunberg16-urges-leaders-to-act-on-climate

1 Climate Change
Here-Now-You-Hope

Hope is not the conviction that something will turn out well, but the certainty that something is worth doing no matter how it turns out.

—Václav Havel

What This Book Is About

Climate change is about you, here and now, and, yes, there is hope.

A few years ago at a senior center in Connecticut, I gave a talk about global warming. I thought the talk went well, and afterward as I stood by the exit an older woman came by and congratulated me on the presentation: "Oh, Dr. Moody, what you said was so powerful. I am just glad I will not be there to see it."

I was taken aback by her comment, but I didn't know what to say. Since that day I've thought about the conversation and I've realized I could have said to her, "No, you will live to see it. You **have** lived to see it. It's already here, all around us." I could have gone further and been more critical. Do any of us really have the right to be happy no longer to see what our hands have left to those who will come after us?

Climate change is already here and now, and, yes, it's about you and, yes, there is hope. There are things we can do about it, both to reduce its impact and to adapt to it as we grow older. Both mitigation—preventing it from getting worse—and adaptation—coping with climate change as it comes—are part of the story. That's the reason there is hope. You didn't notice what was happening? It all happened as you were growing older. Sometimes, the facts have been staring us in the face.

We can debate whether climate change is "the only story that matters."[1] But, for those of us now growing older, climate change has turned out to be the big story of our own old age, even if we never expected it to be. And not only for us: it will be the big story for our children, and their children, and so on.

Climate change is happening at the same historical time when the population of America—and of all industrialized countries—is going through the unprecedented transition of population aging. In the USA, by the end of this

DOI: 10.4324/9781003345992-1

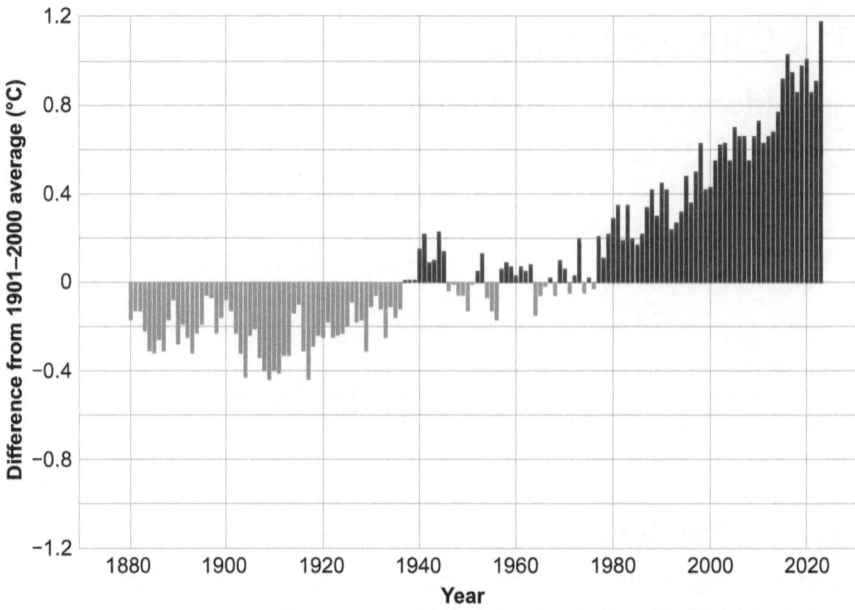

Figure 1.1 Global Average Surface Temperature. Yearly surface temperature from 1880 to 2023 compared to the 20th-century average (1901–2000). NOAA Climate.gov graph, based on data from the National Centers for Environmental Information.

decade, the American population overall will be 20% over the age of 65—or what the State of Florida looks like today. Each day, 10,000 Boomers turn 65. On a global basis, by 2018 the world reached a point where, for the first time in history, there were more people over 65 than children under five years of age. By the year 2030, all the Baby Boomers will be 65 or older. The Age Wave is upon us.[2]

The U.S. National Oceanic and Atmospheric Administration[3] estimates that sea level in the next hundred years will rise by one foot—as much as the increase in the prior century. In other words, sea level rise has already been underway for a hundred years and will increase during the lifetime of everyone alive today. Nearly 40% of the U.S. population lives in coastal areas: in short, HERE NOW YOU.

The current challenge of climate change poses a degree of risk and resilience no one alive has experienced except during wartime or in brief natural disasters. Those who are afraid of population aging—and there are still some with that view—underestimate two converging trends. Population aging is occurring at exactly the time we are facing a "gray rhino"—an event demanding action but an obvious danger that we ignore:[4] Obvious but unforeseen, we might say.

At the end of *King Lear*, William Shakespeare's definitive play about aging, the character Edgar sums up the tragic events that unfolded in these words: "The oldest hath borne most; we that are young/Shall never see so much nor live so long." About climate change, we must say something different: "The oldest hath borne the most, Those that are young shall bear their burden and even more than that."

Climate change isn't just about facts. It's about feelings, above all. "How do you feel about climate, isn't a rhetorical question?" The question asks us to look in the mirror, whatever our age. Kimberly Nicholas describes the five stages of climate feelings.[5] In the first stage, we need to move from (1) *ignorance*, which is where we all start. As ignorance gets chipped away, we're tempted by (2) *avoidance*—otherwise, we feel overwhelmed, falling into climate anxiety.[6] The risk is falling down the black hole of (3) *doom* and despair, even doomerism.[7] Equally bad is the opposite reaction. If cognitive dissonance takes over, we might end up with feelings of self-righteousness and be angry at those who are complacent. (4) *All the feelings*, positive or negative, are necessary. We can't escape any of them to be resilient in surfing the waves of the climate crisis. The key is to face our feelings and harness their strength for action. Be Here Now, as Ram Dass urged in his book on spiritual development. Climate is about you, about me, about us. We succeed, whether young or old, when our feelings point us toward (5) *purpose*, toward hope. HERE NOW YOU HOPE, the watchword of this book.

Two big stories—population aging and climate change—both are happening at the same time in history. We all know climate change is not a happy subject. Neither is aging a happy subject. It's not surprising that we often try to avoid both. Both climate change and aging demand facing up to limits but facing limits is not the American way. On an individual level, a lot of energy goes into avoiding aging: cosmetics, plastic surgery, exercise at the gym, diet, supplements, and all the rest.

Some of it works. Much of it doesn't. But, whatever the result, most of us recognize that we can't avoid growing older, and we know it. Or to put it differently: we know it but we don't face it. I've worked in the field of aging for 50 years, and on my 77th birthday I received as a gift a sweatshirt saying, "I thought growing old would take longer." The birthday gift confirmed what I already had been believing. I was obviously mistaken.

Denial is deep, and aging takes place without our noticing it: just like climate change. As for climate change, well, we often think, maybe it's something for the future, or maybe for people in some vulnerable places on the earth. The psychologist Per Espen Stoknes, in his book titled *What We Think About When We Try Not to Think About Global Warming*, puts his finger on the problem:

> Climate change is perceived as happening elsewhere: the Maldives, Philippines, Arctic, or Antarctic, or in New Orleans or upper Himalayan valleys or the Bangladesh delta. It's about "them," not "us."

The evidence of climate change is all around us, but we do not see—we do not call out:

> When springtime comes too early
> When bees and butterflies disappear
> When California smoke comes over New York
> When atmospheric rivers drench our streets
> When fireflies from childhood go away
> Slowly, slowly, night thoughts come along:
> What has gone wrong with our world?
> What is it we must do?
> The Elders remember, but will they call out?

The Gospel of Mark urges hope: "Blessed are your eyes because they see, and your ears because they hear" (Mark 13:16). The blessing comes with crying out, which is a call for hope.

Day after day, we think, maybe it will happen in the future but not too soon. Just like my sweatshirt about growing older, we can say, about climate as well as about aging: "I thought climate change would take longer." But short-term thinking is a barrier to climate response: "That is, until your own house (almost) burns down, as Californians in increasingly wildfire-prone areas are experiencing. The violent crackling is what global warming sounds like, and the odor of the noxious smoke is what it smells like when you're close to it."[8]

Samuel Johnson put it best: The prospect of being hanged tomorrow concentrates the mind wonderfully.

Short-term thinking, about aging and climate change, is deep, hard to see, and hard to challenge. We're all stuck in old mind patterns, and so it's natural that we don't prioritize long-term climate considerations.[9] Psychologist and economist Daniel Kahneman offers insight into cognitive psychology about preventing action on climate change.[10] As Kahneman puts it, thinking fast orients us to give a short-term response to problems. But climate change demands thinking slowly, thinking long-term. Climate change "makes us feel *helpless* since—even if we stopped emitting now, they say—its delayed effects (including our grandfather's coal burning from the last century) will continue trouble us in decades and centuries ahead."[11] No wonder it's easy to lose hope and easy to turn our eyes away.

This book takes a different view. The message of the book can be summed up in four words: YOU HERE NOW HOPE. It's about YOU, whatever your age today, if you hope to grow older. Climate change is not some distant threat; it's already HERE.

Climate change is NOW, already affecting you and friends and family, everyone you know, and everyone you don't know. You see it today, and you'll see it in 2025, 2030, 2040, and in years after that and again after that. The last word is HOPE. So where will you find it? Not by looking away, not by denial.

We have good reason to fear for the future, our own future, and our children and grandchildren. Surveys show that Americans are becoming more pessimistic about the future.[12] We should have fear because we stand at a crossroads, well described in the American indigenous tradition as the "Seventh Fire Prophecy":

> The Seventh Fire Prophecy presents a second vision for the time that is upon us. It tells that all the people of the earth will see that the path ahead is divided. They must make a choice in their path to the future. One of the roads is soft and green with new grass. You could walk barefoot there. The other path is scorched black, hard; the cinders would cut your feet. If the people choose the grassy path, then life will be sustained. But if they choose the cinder path, the damage they have wrought upon the earth will turn against them and bring suffering and death to earth's people.
>
> We do indeed stand at the crossroads. Scientific evidence tells us we are close to the tipping point of climate change, the end of fossil fuels, the beginning of resource depletion. Ecologists estimate that we would need seven planets to sustain the lifeways we have created . . . Whether or not we want to admit it, we have a choice ahead, a crossroads.[13]

As Mariame Kaba put it: "Hope is not optimism. Hope is a discipline . . . we have to practice it every single day."[14] Ram Dass constantly said Be Here Now, encouraging us to look within to find out how to act in the outer world: "We are never far from the answer to the problem we have created—it is within each of us."[15]

Indigenous traditions are understood to believe that, in every deliberation, we must consider the impact on the seventh generation. Should elders—should all of us—be thinking about Seven Generations? One answer is clear enough:

> A full scale, all hands on deck response to the climate crisis will be successful within 7 generations. Arrogance is thinking that success will be accomplished in our lifetimes. But without that full scale response, there will not be a seventh generation.[16]

If your response (like mine) is "Yes," then look to the past and prepare for tomorrow, as in the science fiction vision of Kim Stanley Robinson:[17]

> "Even as the best-case scenario slips through our fingers, the worst-case scenarios are off the table because of the decline of coal, the rise of renewables, the spread of electric vehicles," said Michael Coren ("The Climate Coach").[18] The chair of the Intergovernmental Panel on Climate Change (IPCC) put it in the starkest terms:[19] "The choices we make now and in the next few years will reverberate around the world

for hundreds, even thousands, of years." Those of us today, whatever our age, are called upon to be Good Ancestors, to care for those who will come after us.

Here-Now-You-Hope. It's a simple message and a short one. Begin with what I will call the Four Horsemen of the Climate Apocalypse (fire, flood, drought, heat wave).[20] The Greek word *apokalupsis* originally means "uncovering," which is why the last book of the Bible is translated into English as the Book of Revelations. That book includes Four Horsemen (pestilence, war, famine, death), which are already at hand: fire, flood, drought, heat wave, the first chapters of this book. Rightly seen, these four are an uncovering, a revelation of what is already there. A 2022 survey by the Pew Research Center found that 71% of Americans reported experiencing at least one of them just in recent years alone. Among the same people, 80% reported that climate change was a factor. These figures coincide with another 2022 survey: more than three-quarters of Americans have been affected by extreme weather events in the past five years. All the survey data reinforces a key point: climate change is not just for the future. It is here and now.

We're all familiar with political polarization about climate change: Democrats mostly see that it's happening, while some proportion of Republicans are still in denial. A big reason for polarization is that the right wing sees climate change as a way for big government to limit people's choices. Those who fear regulation will be reluctant to face the threats around us. This point underscores the virtues of the 2022 Inflation Reduction Act. Instead of regulating or punishing people for actions damaging to the climate, the law provides incentives and rewards (e.g., tax credits) for adopting steps that reduce greenhouse gas emissions. Instead of punishment and sacrifice, we need language that emphasizes human flourishing in a post-fossil-fuel age.[21] We should not avoid genuine fear. But speaking to people in the more hopeful language of human flourishing can help build the will needed to implement climate solutions already available.

That woman at the senior center who told me she expected to die before climate change happened was not in denial. Not at all. She heard the message loud and clear. That's why she felt relieved to know that she would be gone before seeing it: that's the opposite of denial. She was expressing something else: not denial, but despair, the belief that nothing could be done. Call it complacency, passivity, paralysis, whatever word you use. It amounts to the opposite of hope. My other big mistake in the talk that I gave was to present such a gloomy negative picture that her conviction of despair was only reinforced. It's the same big mistake that many environmental advocates have made in talking about climate change. Scare people and maybe they'll change. But that's not an approach which works. Human beings are hardwired to pay attention to bad news. It begins in infancy, and it's accentuated by both social media and mainstream media: more clicks, more attention in the "attention economy."[22] There is an alternative on many subjects, including climate change.

Where is the path of hope?

It's true that we still have climate deniers among us, but their number has dwindled, according to the most reliable surveys. Extreme weather events have helped with persuasion. We see climate change all around us.

The book you're reading is not written for climate deniers: that is, people who simply don't want to face what's already happening: fire, flood, heat waves, and drought. This book is based on HOPE, the last of those four words.[23] We can do something to reduce the speed of global warming (Mitigation), and we can cope with the impact of what happens to us (Adaptation).[24] That's why we have hope. It's about doing and coping. As environmentalist David Orr put it: "Hope is a verb with it is sleeves rolled up." The aspiration in a time of dangerous climate change is best expressed by psychologist Erich Fromm:

> Only through full awareness of the danger to life can this potential be mobilized for action capable of bringing about drastic changes in our way of organizing society . . . One cannot think in terms of percentages or probabilities as long as there is a real possibility—even a slight one—that life will prevail.[25]

The book you're reading has a restricted focus: climate change and aging. It's restricted geographically, too, and restricted in time. Attention is on the USA, even though residents of the Global North can learn things from those in the Global South who are already moving toward adaptation in ways that we have not imagined. For example, the death toll from flooding in Bangladesh[26] exceeds what we in the USA have seen even from "atmospheric rivers." But even in adverse circumstances, there is cause for adaptation and thus for hope.[27]

In this book, we look primarily at American responses and we look to the immediate future: the current decade of the 2020s, when decisions today will shape our world far into the future.[28] Restricted focus means there will not be enough time or attention to all these threats except where the threats are clearly linked to the climate crisis. Here and now: that's a limit, but, like a concentrated laser beam, it can focus our attention on what we, as individuals and acting with others, can do to reduce the threat that is at hand.

In the decade of the 2020s, we are living at a time that some have called a polycrisis.[29] It is a time when many existential threats are converging: biodiversity loss, resource depletion, plastic pollution, ocean acidification, economic inequality, and more. In this book, I focus exclusively on climate change, and there's a reason for this focus. The reason is that extreme weather events are what people experience here and now: one of the primary themes of this book. Other disasters happen "somewhere else" (i.e., outside the USA) and are therefore ignored, unless they reach catastrophic levels, as happened with flooding in Pakistan and some other cases. But our human tendency is to ignore what is not visible, not tangible. The point holds true for the ways in which climate change yields temporary benefits, such as immense wealth for the few, easy mobility, cheap food, a burgeoning middle class, and a wide variety of consumer goods.[30]

All these temporary benefits are linked to the exploitation of fossil fuels. And many of us who are not poor gain these temporary benefits. It is indisputable about what needs to be done: reduce carbon emissions. But the latest reports from the Intergovernmental Panel on Climate Change confirm that carbon emissions are increasing and have been for decades, despite promises and efforts. How many of us will be protesting about air conditioning, living in concrete buildings, flying in airplanes, or using plastic materials in our homes? Temporary benefits are taken for granted, especially in the Global North. My point is not to condemn our appreciation of temporary benefits but simply to acknowledge the fact. It underscores why attention in this book is not on all the existential threats but only on those that are tangible: HERE NOW YOU.

The present book is concerned with aging, make no mistake about it. Note here the active present participle: aging-**ing**. Aging is a process, and it affects people of different chronological ages. It's not about people who happen to be old today, like the author of this book (age 79). The book adopts a life-course perspective on aging where old age is viewed over the complete trajectory of life. There are those who are already old today, such as the author. And there are now those who are "not-yet-old" but have hopes of reaching old age. That means everyone reading this book. The young and middle aged are preparing now for what their own later life will be, like it or not. Those now old are best thought of as "our future selves."

This book highlights big problems but above all looks for solutions. Where do we find hope? How do we "roll up our sleeves?" Again, there is a restricted focus: individual agency. We're trying to answer a question: What can one person do?

The group Green America offers ten answers to that question.[31] Their recommendations are parallel to what is discussed in this book. For example, in the chapter on "Becoming a Climate-Conscious Consumer" we consider Eliminating Food Waste; Eating Plant-Based Food; Using Clean Energy; Using LED Lighting; Improving Home Insulation; and Recycling. Above all, Buying Less, where "Refuse" is the fourth "R" added to Reduce, Reuse, Recycle. In the chapter on "Investment for a Green Retirement," we look at the divestment of savings from fossil-fuel investments. In the chapter on "Citizen Climate Action" we will look at rethinking transportation and, above all, participating as a citizen at all levels. "Think globally, act locally" remains the watchword for action.

It is easy for elders to fall into despair. The great psychologist Erik Erikson told us that aging, the last stage of life, is dominated by the polarity of ego-integrity versus despair. The climate crisis displays it all around us. Jane Goodall said, "Here we are, arguably the most intelligent being that ever walked planet Earth, with this extraordinary brain, yet we're destroying the only home we have."

1.1 Jane Goodall: Eco-Elder[32]

Figure 1.2 Jane Goodall. Photography by Muhammad Mahdi Karim.

I had the great good fortune to meet Jane Goodall in person at the 2010 AARP National Conference, when I introduced her as she gave a major address to AARP members. Speaking to thousands of elders she gave a message she has repeated in the years since then: "I just know that if we carry on with business as usual, we're going to destroy ourselves. It would be the end of us, as well as life on Earth as we know it," warned Jane Goodall.

She articulated her message with clear implications for what would later be the COVID-19 pandemic. She pointed to the role of zoonotic diseases that leap from animals to humans, noting that biologists have long been predicting this. In Wuhan or Africa or the Amazon, we are chopping down the rainforest and coming into contact with animals. By eroding biodiversity, we end up creating environments where microorganisms can cross species, as COVID-19 has evidently done. Jane Goodall has worried that the climate crisis is being put into second place because our attention is riveted by the pandemic.

Since the time she spoke at the AARP conference, Jane Goodall has turned 80 and she travels more than 300 days a year, bringing her warning to groups around the world. But she is much more than a prophetess. She has created connections across the generations.

In 1991 Goodall launched a program for young people known as Roots and Shoots. All over the world, she met young people who had lost hope in the future. In her eyes, it is the role of elders to recognize that "we have compromised their future and there was nothing they could do about it." Jane Goodall's answer is that it is not too late, that elders and young people can work together to slow down climate change and educate the world.

In her own words: "I have to work with young people today so that we try and raise new generations to look after this poor old planet better than we have, before it's too late."

Erik Erikson said that the last stage of life is framed by the struggle between ego-integrity and despair. Jane Goodall works to overcome despair. This is the path of action and the path of healing, and, into her eighties, Jane Goodall does not stop working for that goal.

But Jane Goodall's response has not been despair. She reminds us that we should never doubt that we can have some impact. As she put it: "You cannot get through a single day without having an impact on the world around you. What you do makes a difference, and you have to decide what kind of difference you want to make."

Jane Goodall's stance of hope corresponds to activism:

I see humanity at the mouth of a very, very long dark tunnel. And right at the end of that tunnel is a star. And that's hope. But it's no good sitting at the end of the tunnel and hoping that star will come. No, we've got to roll up our sleeves, climb over, roll under and work around all the obstacles that lie between us and the star. Like climate change, loss of biodiversity, poverty where people destroy the environment to survive, overconsumption. We've got to work on all these things to reach the star. But good news: there's people working on every single one of them.[33]

We can add to Jane Goodall's statement the point that none of us is acting alone. Individual acts can influence collective decisions by business and government. Some simple steps, including consumer choices, fall within the scope of individual behavior. Others depend on policy decisions at the community, state, or national level. When we're asked what can only one person do, the answer is: Don't just act as one person.

Every single reader of these words is growing older during this time of climate change. It's a mistake to think of individual action in opposition to collective action. Putting it that way is a false alternative. Both are indispensable, and both are connected. Individual acts are always limited: my single act of voting hardly ever decides a single election. But my individual act is tied to what others do outside the polling both, before and after an election. Whenever people feel "The future can't be changed" or "Individuals can't a difference," they're already on the path to despair and passivity. This book is an extended answer to the question, "What can one person do?"

The response can be summed up in four words on the first page of its website:[34] YOU HERE NOW HOPE.

True, discussion is restricted to climate and aging, and other important topics are left out. But, like a laser beam, when focusing restricted energy this way, we get maximum impact. We get a picture that is both realistic and hopeful, a picture that permits each of us to act on behalf of the world we want for ourselves and for future generations.

There is no better example of what I call an Eco-Elder—elders acting on behalf of future generations—than David Attenborough, the oldest of the Eco-Elders cited in this book.

1.2 David Attenborough: Eco-Elder

We are facing a man-made disaster on a global scale. Our greatest threat in thousands of years.

—Sir David Attenborough

Sir David Attenborough, now age 96, has long been renowned for his books and broadcasts celebrating the wonders of the natural world. But in later years he has been known for his support of environmental causes, such as biodiversity, renewable energy, and advocacy around climate change. He has been recognized by being given the Champion of the Earth award from the United Nations. His words in old age are a message for all generations:

The scientific evidence is that if we have not taken dramatic action within the next decade, we could face irreversible damage to the natural world and the collapse of our societies. We are running out of time, but there is still hope.

(David Attenborough) September 17, 2022

Life Review: A Planet in Crisis

"Life review" is a term invented by geriatrician Robert N. Butler, M.D., in an influential article he wrote in 1963. Butler there argued that we should not dismiss the reminiscence of older people but instead should see the process of recollection in therapeutic terms. Reviewing one's life could offer elders positive reconciliation through recalling the events of a lifetime.

I had the great good fortune to know Dr. Butler personally and to work for him at the International Longevity Center in New York. He became a mentor for me, and his invocation of the positive dimension of later-life memory is inspiring. He specifically encouraged me to address the ethical challenge of justice between generations, which is at the heart of the book you are reading.

Dr. Butler's concept of life review leaves open a question. What about those who have lived long enough that historical experience challenges their past and forces one to ask painful questions about our collective world? What about the grim facts that threaten our sense of hope?

For those concerned with life review and climate change, the film by David Attenborough *A Life on Our Planet* is a must-see experience. The film follows Attenborough as he traces his own life history from boyhood to becoming world-renowned naturalist. The film, in essence, is a work of life review in the fashion that Robert Butler imagined. The film is also what Attenborough himself calls a "witness statement." It's the voice of one who has witnessed wholesale destruction of biodiversity on a global scale. How to reconcile his gratitude for good fortune with the dismay near the end of his life, as he sees what will be happening in decades to come? To his credit, Attenborough never flinches, never looks away from the climate crisis. But he also ends the film with hope and with a call for action, in his words "while there is still time."

Writing this book has also been for me an exercise in life review. From the time I was given my Eagle Scout Award in 1960 until the present, I have seen, like others of my generation, how the planet has altered in ways I never would have expected. The period of climate change in the decade of the 2020s is not the retirement time I had imagined. Like David Attenborough's "witness statement," I have written a letter of apology to my two granddaughters, who would be my age in the decade of 2090s.

Like many others of my age, I ask: What kind of legacy are we leaving to those who come after us? I once was sitting in a restaurant with Rabbi Zalman Schachter, talking with him about the search for meaning in old age. Suddenly Reb Zalman turned to me and said, "You know, it all comes down to a simple question: Are you saved?"

Hearing what he said, I was puzzled, and I thought, well, he is a clergyman after all: "Is this perhaps a theological reference?" I asked. "No," said Reb Zalman, "I don't mean it in a theological sense but in a computer sense. Are you saved? Have you downloaded your life experience for the coming generations? Are you doing your legacy work?"

1.3 Zalman Schachter: Eco-Elder

Figure 1.3 Zalman Schachter. Used with permission of the estate of Zalman Schachter.

Zalman's challenge "Are you saved?" raises the question: What kind of world will we be leaving to our children and grandchildren? What is the collective legacy of our time on this earth? Reb Zalman spoke of this challenge as a spiritual task, the task of "outliving the self," or what he came to call "sage-ing."[35]

In earlier years Reb Zalman founded the movement of New Jewish Renewal. But after he turned 60, the rabbi was troubled by what old age might be for him. What is the purpose of these extra years of life? I got to know him in Boulder, Colorado, where I lived for some years. I saw how Reb Zalman devoted the remaining years of his life to creating a new vision of what later life could be: in his own words, "From Age-ing to Sage-ing." His concern for environmental issues reflected an abiding conviction about the purpose of the later years of life.

When it comes to climate change, Reb Zalman answered, yes, elders can save the world.[36]

What is "legacy work?" Some good answers are available in *The Good Ancestor*, by Roman Krznaric.[37] But to hear the answers we need to want to hear, not to close our ears to the warnings that are all around us.

When people think about climate change, the problem is NOT denial. The problem is passivity, complacency, and helplessness. It's that little voice which says, "What can one person do?" Plenty of people already know there's a problem. But pessimism or learned helplessness—"What can I do?"—intervenes, and so nothing happens.[38] "Too apathetic." The big adversaries for stopping the climate threat? Complacency, passivity, and learned helplessness. That fantasy that "someone else will do something." Time to wake up? We are not helpless to act, even if a single act doesn't change the world.

"Learned helplessness" is how gerontologists describe what happens to older people with limited capacities who enter a long-term care facility. They acquire—that is, they learn—how *not* to assert themselves or confront the reality around them. This happens with climate change, too. But in climate change this takes the form, not of institutional power as in long-term care but the power of a "demonic voice" or what Buddhists call the "daughters of Mara."

Mara is the king who tried to seduce Buddha away from the path of enlightenment. Two of his demonic daughters are fear and pride or what we today might simply call cowardice: the demonic voice that whispers to politicians never to acknowledge climate change, even when extreme weather events prove before their very eyes what the demons are saying is false. But cowardice shuts their mouth so the politicians never say out loud what they know in their hearts. They, too, have gone down the path of learned helplessness. Facts will not change their minds. Their problem is not denial but falling under the power of Mara's daughters called fear and pride, who together keep political leaders in cowardice.

The threat today is what Robert Jay Lifton called "psychic numbing." I met Prof Lifton only once, when we were both in a group to receive awards from the City University of New York, where I worked for many years. Lifton, who has taught us as much about trauma as anyone on the earth, could be called a National Treasure, as they say in Japan. What Lifton called "psychic numbing" is a diminished capacity or inclination to feel. One point about psychic numbing, which could otherwise resemble other defense mechanisms, like de-realization or repression: it is only concerned with feeling and nonfeeling.[39]

The Greeks understood the world in terms of four elements: earth, air, fire, and water. In the case of Troy, Cassandra warned the Trojans about their peril from the Greeks, but they rejected her and allowed the horse to come into their city. Then, at nighttime, in darkness, the Greeks sprang out of that horse and destroyed the city of Troy. In our time engineers use the term "Trojan horse" to describe deceptively benign computer codes that look like legitimate software but are designed to damage or disrupt a computer's operation. In the case of our current civilization, most of our energy comes from fossil fuels—coal, oil, and gas—which are carbon compounds from once-living plants, sunshine buried under the earth, invisible for most of human history. In the blink of an eye—only two centuries—these carbon compounds came out from the Shadow World beneath the earth. They have been brought to the surface to create fire.

Wind and water cannot stop this fire, and the fire burns and produces an invisible, undetected pollutant, carbon dioxide, accumulating in the air until ice is melted and turns water into a destructive force. The Shadow beneath the earth has become a new Shadow across the surface of the globe. It is also a Shadow within ourselves, but we turn our eyes away from it. Can we assemble strength and wisdom to resist the daughters of Mara, to hold back the Trojan Horse, to escape the Shadow that threatens our very lives? We who are older find ourselves slipping into learned helplessness, which takes the form of learned hope-lessness—unless we find the strength and wisdom to resist.

Václav Havel, who risked his life to overturn the Communist dictatorship in Czechoslovakia, put it best when he said that hope is not the conviction that something will turn out well but the certainty that something is worth doing, no matter how it turns out. That's our situation when it comes to climate change. To find that strength and wisdom can be summed up in three questions: What do we know? What should we do? What can we hope for? Each question must be personal, and each question for you, as reader, must include an answer with these four words: HERE NOW YOU HOPE.

This last point—HOPE—is where action begins. Anyone who reads the first two parts of this book will get a horrifying glimpse of the Four Horsemen of the Climate Apocalypse: fire, flood, heat wave, and drought. These are not threats for someone else, for another part of the globe. Not at all. One or another of the Four Horsemen will come for you, the one reading these words, or someone you love and care about. It might not happen today or tomorrow. But the threats will come. It's not for the future. As one will put it, the future has already arrived: it's just unevenly distributed.

Immanuel Kant (1724–1804) was arguably the greatest philosopher of modern times. When Kant became an old man, he was asked, "What is the essence of philosophy?" His reply came down to three questions: What do we know? What should we do? And what may we hope? In this book we begin by recognizing what we know: the Four Horsemen of the Climate Apocalypse: fire, flood, heat wave, and drought. These are the threats we experience directly in our bodies, and this is what we know. We then move on to the ethical challenge: What should we do? This second question is the question of ethics: recognizing who is to blame and acknowledging what we must do in response. Actions are whatever is in our power, and in this book we look at the right steps of climate ethics: becoming a climate-conscious consumer; investing in a green retirement; and acting as citizens so that the government will respond to climate change. Finally, what may we hope? We turn to both hope and fear and consider why we need both hope and fear to act in the time we have available, the time we are living in now: HERE NOW YOU HOPE.

Notes

1 Charles Pierce, "In the End, Climate Change Is the Only Story That Matters" *Esquire* (Sept. 24, 2022) at: www.esquire.com/news-politics/politics/a41355745/hurricane-fiona-climate-change/

2 U.S. Census Bureau, "By 2030, All Baby Boomers Will Be Age 65 or Older" at: www.census.gov/library/stories/2019/12/by-2030-all-baby-boomers-will-be-age-65-or-older.html See also: Lawrence MacDonald, *Am I Too Old to Save the Planet? A Boomer's Guide to Climate Action* (Changemaker Books, 2023).

3 NOAA at: https://oceanservice.noaa.gov/hazards/sealevelrise/sealevelrise-tech-report.html#:~:text=Sea%20level%20along%20the%20U.S.,years%20(1920%20-%202020)

4 Vinod Thomas, "Risk and Resilience in the Era of Climate Change" *Brookings* (Apr. 4, 2023) at: www.brookings.edu/blog/future-development/2023/04/04/risk-and-resilience-in-the-era-of-climate-change
      See also Michele Wucker, *The Gray Rhino: How to Recognize and Act on the Obvious Dangers We Ignore* (Saint Martin's Press, 2016).

5 Kimberly Nicholas, a climate scientist, puts it in front of us: *Under the Sky We Make* (Putnam, 2021). See also K. Nichols, "5 Stages of Climate Feelings" *We Can Fix It* (Aug. 31, 2023) at: https://wecanfixit.substack.com/p/5-stages-of-climate-feelings

6 "[Eco-anxiety] sensitizes individuals to situations where they face novel or difficult decisions about something of ecological value; and it prompts the cognitive engagement and motivation that can help them address the difficulty they face. Thus, eco-anxiety is an emotion that brings forms of awareness and engagement that can contribute positively to environmental stewardship and agency as well as individual and collective well-being" (From the Carbon Almanac Network).
      See also "How to Manage Eco-Anxiety" in *Gray Is Green* at: https://mailchi.mp/9f20bd298925/earth-month-online-5415643?e=a1f7f673d8

7 Doomerism may actually obstruct our ability to act or to have hope. The attitude of "Why bother? It's hopeless" doesn't prompt people to do something. Such pessimism may not be necessary at all: "What If People Don't Need to Care About Climate Change to Fix It?" at: www.nytimes.com/interactive/2023/12/31/magazine/hannah-richie-interview.html
      See also: Hannah Richie, *Not the End of the World: How We Can Be the First Generation to Build a Sustainable Planet* (Little, Brown, 2024).

8 Per Espen Stoknes, *What We Think About When We Try Not to Think About Global Warming* (Chealsea Green, 1915), pp. 29, 32.

9 Ibid., p. 34.

10 Daniel Kahnemann, *Thinking Fast and Slow* (Farrar, Straus and Giroux, 2011).

11 Stoknes, *What We Think About When We Try Not to Think About Global Warming*, p. 40.

12 Andrew Daniler, "Americans Take a Dim View of the Nation's Future, Look More Positively at the Past" *Pew Research Center* (Apr. 24, 2023) at: www.pewresearch.org/short-reads/2023/04/24/americans-take-a-dim-view-of-the-nations-future-look-more-positively-at-the-past/

13 Robin Wall Kimmerer, *Braiding Sweetgrass: Indigenous Wisdom, Scientific Knowledge and the Teachings of Plants* (Milkweed Editions, 2013). For more on the contribution of Indigenous wisdom see "The Climate Crisis and Aging: Capitalizing on Traditional Knowledge and Innovation" at: www.justsecurity.org/90762/the-climate-crisis-and-aging-capitalizing-on-traditional-knowledge-and-innovation/

14 https://climategen.org/blog/hope-is-a-discipline/

15 From Tara Hourska, *Sacred Resistance*, cited in *All We Can Save: Truth, Courage, and Solutions for the Climate Crisis*, edited by Ayana Elizabeth Johnson and Katharine K. Wilkinson (One World, 2021).

16 These are words from Robert Whitehair, an Eco-Elder: www.youtube.com/watch?v=GRu_FwmidlU&t=10s

17 https://longnow.org/ideas/in-the-ministry-for-the-future-new-ideas-from-ancient-wisdom/ See Kim Stanley Robins, *Ministry for the Future* (Orbit, 2021).

18 Michael J. Coren, "The Climate Coach" *Washington Post* (Mar. 23, 2023).

19 "World is on brink of catastrophic warming, U.N. climate change report says" A dangerous climate threshold is near, but 'it does not mean we are doomed' if swift action is taken, scientists say at: www.washingtonpost.com/climate-environment/2023/03/20/climate-change-ipcc-report-15/?utm_source=newsletter&utm_medium=email&utm_campaign=wp_climatecoach&wpisrc=nl_climatecoach

20 The language of climate and apocalypse is hardly original here. See Vítězslav Krmlik, *A Guide to the Climate Apocalypse: Our Journey from the Age of Prosperity to the Era of Environmental Grief* (Identify Publications, 2021).

21 Susan Hassol, "The Right Words Are Crucial to Solving Climate Change" *Scientific American* (Feb. 1, 2023), pp. 64–67 at: www.scientificamerican.com/article/the-right-words-are-crucial-to-solving-climate-change/

22 Thomas Davenpport and John Beck, *The Attention Economy: Understanding the New Currency of Business* (Harvard Business Review Press, 2001).

23 On hope and optimism, see Terry Eagleton, *Hope Without Optimism* (University of Virginia Press, 2021).

24 On adaptation, see The Climate Disaster Project at: https://climatedisasterproject.com/
   And Carbon Almanac Network at: dailydifference@thecarbonalmanac.org
   See also Jem Bendell and Rupert Read (eds.), *Deep Adaptation: Navigating the Realities of Climate Chaos* (Polity, 2021).

25 Erich Fromm, *The Revolution of Hope: Toward a Humanized Technology* (HarperCollins, 1981).

26 Zoya Teirstein, "The Hidden Death Toll of Flooding in Bangladesh Sends a Grim Signal about Climate and Health" *Grist* (Dec. 11, 2023) at: https://grist.org/health/the-hidden-death-toll-of-flooding-in-bangladesh-sends-a-grim-signal-about-climate-and-health/

27 Mongabay.com, "'Adaptation Bangladesh: Sea Level Rise' Film Shows How Farmers Are Fighting Climate Change" (Feb. 22, 2018) at: https://news.mongabay.com/2018/02/adaptation-bangladesh-sea-level-rise-film-shows-how-farmers-are-fighting-climate-change/

28 See the "12 Biggest Environmental Problems of 2022" at Earth.Org at: https://earth.org/the-biggest-environmental-problems-of-our-lifetime/?mc_cid=b5a0cdace7&mc_eid=50625bb884

29 Adam Tooze, "Welcome to the World of the Polycrisis" *Financial Times* (Oct. 28, 2022). See also:
   Adam Tooze, "Polycrisis—Thinking on the Tightrope" (Oct. 22, 2022) at: https://adamtooze.substack.com/p/chartbook-165-polycrisis-thinking

30 My appreciation here to Richard Heinberg, "Looking Back, Looking Forward" *PostCarbon Institute Newsletter, "Messenger"* (Mar., 2023). For more see Richard Heinberg, *Power: Limits and Prospects for Human Survival* (New Society, 2021).

31 Mary Meade, "10 Ways You Can Fight Climate Change" *Green America* at: www.greenamerica.org/your-green-life/10-ways-you-can-fight-climate-change

32 Through the rest of this book I will highlight certain individuals as "Eco-Elders." Just as Japan designates certain buildings and structures as "National Treasures" so we should designate certain individuals for their exemplary work in prizing planet Earth. The United Nations now does exactly this by means of its "Champions of the Earth" award: www.unep.org/championsofearth/laureates/2023
   Eco-Elders cited are worth celebrating and are included periodically in this book.

33 McMoult Craig, "Despite Environmental Challenges, Jane Goodall Says There Are Reasons for Hope" *GBH* (Oct. 3, 2023) at: www.wgbh.org/culture/film-and-tv/2023-10-03/despite-environmental-challenges-jane-goodall-says-there-are-reasons-for-hope

34 www.climateandaging.org

35 John Kotre, *Outliving the Self: Generativity and the Interpretation of Lives* (Johns Hopkins University Press, 1984) and Zalman Schachter-Shalomi, *From Age-Ing to Sage-Ing: A Revolutionary Approach to Growing Older* (Time-Warner, 1997).

36 "Can Elders Save the World?" by Rabbi Zalman Schachter-Shalomi at www.yes-magazine.org/issue/elders/2005/08/30/can-elders-save-the-world

   From Yes Magazine, "Honoring Our Elders" at: www.yesmagazine.org/issue/elders-2/2023/11/30/retirement-honor-elder

   See also Thelma Reese and B.J. Kittredge, *How Seniors Are Saving the World: Retirement Activism to the Rescue!* (Rowman & Littlefield, 2020).

37 Roman Krznaric, *The Good Ancestor: A Radical Prescription for Long-term Thinking* (The Experiment, 2021).

38 The original analyst of learned helplessness, Martin Seligman, followed up with *Learned Optimism: How to Change Your Mind and Your Life* (Vintage, 2006). On overcoming learned helplessness, see also John Izzo, *Stepping Up: How Taking Responsibility Changes Everything* (2021). See also: Kimberly Nicholas ("We Can Fix It") "Explaining climate inaction" at: https://wecanfixit.substack.com/p/explaining-climate-inaction

39 From "How to Maintain Hope in an Age of Catastrophe" *The New Yorker* (Nov. 12, 2023) at: www.newyorker.com/news/the-new-yorker-interview/how-to-maintain-hope-in-an-age-of-catastrophe

   Lifton is the author of many important books on "totalism," including *The Climate Swerve: Reflections on Mind, Hope, and Survival* (New Press, 2017).

   But see also: "One Huge Contradiction Is Undoing Our Best Climate Efforts" The math isn't adding up, by Zoë Schlanger *Atlantic Monthly* (Nov. 10, 2023).

Part I

Climate Change and Aging in Place

2 The Fire Next Time
Wildfire

Liar, liar, the world's on fire
What you gonna do when it all burns down?
Fire, fire burning higher
Still got time to turn it all around.

—Dolly Parton, "World on Fire"

Surveys from AARP tell us that up to 90% of older Americans want to age in place and stay right where they are. But anyone reading these words knows perfectly well that we're all living in a time of climate change. If you go to the website of the National Institute on Aging to find out about "Aging in Place," you'll get lots of good guidance about things to do as an individual to make your home safer and more congenial for growing older. What you won't find is even a single reference—not one—to environmental threats like fire and flood, drought, and heat waves—that also affect your ability to age in place. Other guidance on aging in place is much the same.[1] It's as if the home in which you live is not even located in a natural environment at all. It's as if saying "Nothing can harm me. I'm in my own home." But this is a delusion, isn't it?

Think globally but act locally is the message we need to hear again and again:

Climate change is often cast as a global issue. But it's already affecting each of us in our backyards. Why it matters: The more people see the impact on their own lives, the more likely it is that they'll look for things they can do about it—from the candidates and policies they support to the personal changes they can make.[2]

Dolly Parton in her song, "World on Fire," demands that we face the question: "What you gonna do when it all burns down?" She begins with

DOI: 10.4324/9781003345992-3

Figure 2.1 Flames of the Simi Valley Fire, in Southern California. One of many wildfires that have occurred in Southern California over the decades. Public
domain.

fear, but she concludes with an answer of hope: "Still got time to turn it all
around."

We know that wildfires and floods, drought and heat waves, could be in our
future, for me and for you: "Fire, fire burning higher." Extreme weather events
are not just on the horizon: they're coming for us already: HERE NOW YOU.
One kind of HOPE is knowing where the threat will be[3] How do we reconcile
the wish to age in place with climate risk that could make it impossible to stay
where we are? What happens if a fire or flood destroys the home we want to
stay in? These are the questions addressed in this chapter and in the chapter
that follows. We will also consider the challenge of safety: is there insurance to
protect against fire and flood? Are there other ways of searching for safety? We
will try to overcome the learned helplessness that keeps us all in the delusion
that it won't affect us or that it's too late and there's nothing that can be done.

Smoke and Fire

Where There's Smoke, There's Fire.
In August 2020, when my wife and I moved from Boulder, Colorado, to San
Mateo, California, we moved into a monstrous wildfire. The day we drove

over the border from Nevada, the smoke kept getting worse and worse until when we finally reached the San Francisco Bay Area, smoke was darkening the skies even at midday. In the middle of the day our daughter was taking our then 4-year-old granddaughter to nursery school, and the little girl looked up and asked "Mommy, why are we going to school at night?" She was asking the right question. In September 2020, smoke from California wildfires made the sky so dark during daytime that solar power production was reduced by 20% during peak hours, according to the National Center for Atmospheric Research.[4]

What happened then was not an isolated event:

> In 2020, California suffered a devastating wildfire season. More than 9,900 fires burned around 4.3 million acres, more than doubling the previous record, killing 33 people and cost the economy more than $19 billion, according to the University of California Davis. Now, a new study led by researchers at the University of California, Los Angeles (UCLA) has calculated a new impact: the burns released so much carbon dioxide that it canceled out the state's attempts to reduce greenhouse gas (GHG) emissions between 2003 and 2019.[5]

The fires had shattered the illusion of safety, for me and for many others across the country. We all heard, "Where there's smoke there's fire." But when it comes to wildfires, the reverse is also true: Where there's fire, there's smoke, even far from where the fire is happening. Fires in California put out smoke that ended up in New York; fires from British Columbia put out smoke that ended up in Washington, DC.[6]

A popular 1976 film was titled *I'm Mad As Hell and I'm Not Going to Take It Anymore*. Maybe that's the best response to wildfire that threatens my prospects for aging in place. Maybe young and old can join together?

On June 14, 2021, I went to the Golden Gate Bridge in San Francisco to join youth protest marchers sponsored by Sunrise, the national group of climate activists who have promoted the Green New Deal. Hundreds of protesters that day were joined by a smaller number of older people—part of the "35-Plus" group affiliated with Sunrise in the Bay Area. I was part of that group of elders, and I joined the protest, which was itself a final act of civil disobedience at the end of a 266-mile march intended to spur lawmakers to action on climate change.[7] The march culminated at the San Francisco home of House Speaker Nancy Pelosi, the same building where a year later a believer in conspiracy theories would try to kill Speaker Pelosi's husband.

2.1   Jane Fonda: Eco-Elder

Figure 2.2 Jane Fonda. Public domain.

In recent years Jane Fonda has become, not just a celebrity, but an icon for using her voice on behalf of climate advocacy.

There are many things she could be using her time for at age 85.[8] But Jane Fonda is concerned both with positive aging and with care for future generations she will never live to see. Her trajectory is called *Outliving the Self*, in the title of John Kotre's book about generativity. It is a combination we all need to embrace.

Fonda put it this way:

The climate crisis requires collective action on a scale that humanity has never accomplished, and in the face of those odds a sense of hopelessness may occasionally descend. But the antidote to that feeling is to do something. The question is: what? Changing individual lifestyle choices like giving up meat and getting rid of single-use plastic won't cut it when time is not on our side. We need to go further, faster. Instead of changing straws and lightbulbs, we need to focus on changing policy and politicians. We need large numbers of people working together for solutions that work for the climate. Nonviolent civil disobedience can help to mobilize that movement.[9]

She has become an Eco-Elder in her final act in life. "The cure for despair is action."

When I joined the protesters at Golden Gate I happened to meet "Jennifer" (as I'll call her), also, like me, a member of the 35-Plus group of the Sunrise youth climate advocacy group. Jennifer had unique qualifications for being there that day. She was a long-time resident of Paradise, California, the town well north of San Francisco, a town that made history on November 8, 2018, when it was totally engulfed by the Camp Fire. That devastating fire destroyed nearly every home and building in Paradise and killed 85 people. It displaced tens of thousands of residents. The final irony is that Jennifer herself was a forest ranger quite familiar with wildfires. My granddaughter had only smelled smoke in the darkness of midday in San Francisco. Jennifer had been through the inferno itself. She had every reason to be on the march, and she had every reason to say, like the hero in Arthur Miller's "Death of a Salesman," "Attention must be paid!" The world paid attention to Paradise three years earlier.

Fire in Paradise

Can you imagine what it would be like to live someplace your whole life and then one day find that your house and the homes of your neighbors had all burned to the ground? What if at the point of retirement, you moved to a place you always wanted to be and then the place where you planned to spend the rest of your life burned up in a single day? You don't have to imagine any further. Consider the town of Paradise, California.

The 2018 fire in Paradise, California, known as the Camp Fire, was among the deadliest wildfires in a hundred years. At its peak, there were 5,500 firefighters working to gain control of the blaze, which took 17 days to get under complete control. The Camp Fire grew and became the largest fire in California history. It burned over 150,000 acres, destroyed almost 19,000 homes, and killed more than 85 people.

As often happens with wildfires, the aging population was especially vulnerable. In Paradise, a quarter of the population was 65 or older, and a quarter were already disabled, much higher than in the rest of California. As it turned out, in Paradise, at the same time that thousands of residents were crowded onto highways trying to escape, there were many elderly and disabled residents who couldn't even get out of their own homes.[10] There are important questions here about adaptation to disasters. Fewer than half of the 27,000 residents of Paradise had previously signed up to get alerts about threatening wildfires. Even when alert calls were attempted, more than half the calls failed for a variety of different reasons.

Behind all the statistics are human stories of elders whose retirement home turned into something very different from what they had ever imagined:

> After barely getting out of Paradise alive before the Camp Fire turned her town to ash, Patty Saunders, 89, now spends her days and nights in a reclining chair inside the shelter at East Ave Church 16 miles away.

It hurts too much to move. She needs a hip replacement and her legs are swollen. Next to her is a portable commode, and when it's time to go, nurses and volunteers help her up and hold curtains around her to give her some measure of privacy. "Never in my life did I think I would end up in a situation like this, but when it's time to go, you got to go," Saunders said.[11]

Officials in the U.S. Forestry Service try to prevent events like the Camp Fire, but climate change is making the problem much, much worse. Since the 1980s, in California temperatures have been more than two degrees higher than a hundred years ago. The four warmest years in California's record included 2017, a year before the Camp Fire. It was also a time of drought, and that drought has been extended into future years. In light of the impact of global warming on the Sierra snowpack, it was unknown when the drought will end:

> "Nine of the ten years of the most extensive fire activity in the United States have been since the year 2000 and by 2050 it is expected that as much as three times the acreage of western forests will burn as a result of global warming." The fire in Paradise may not be just past but a glimpse of our collective future. We can look at what's already happened in this century to get a picture of that future. The environmental group known as Earth.org has produced a compendium of the 15 largest wildfires in US History.[12]

Here's a picture of the second worst, the Bay Area Fire, that happened the year I moved to California:

> Starting in the Bay Area, the Bay Area fire was one of the largest wildfires in US history and tore through parts of California, Oregon and Washington state. By September 15, they burned almost one million acres of land and killed at least 35 people. At one point, every 24 hours, an area the size of Washington DC was being burned. The North Complex fire alone was responsible for more than 300 000 acres of scorched land, killing 16 people in its wake. Five of the six largest blazes in the state were recorded in 2020. Meanwhile, Stanford researchers estimate that the smoke and resulting poor air quality eventually led to hundreds of excess deaths in California cities and across the west coast in Washington and Oregon.[13]

We know that many Americans—including my own parents in their old age—will migrate to Florida for retirement living. My parents ended up in the path of hurricanes that become more and more severe with climate change, as we saw with Hurricane Ian in 2022. Eventually, the threat was one factor that prompted them to move into a senior living facility. In the case of Paradise,

California, the town itself became a point of attraction for older people with antiques to sell.[14] The dark side of this migration is that three-fourths of those who died in the Camp Fire were 65 and older.[15]

Clearly, we need to recognize the mental health impact of fire on older people:

> We know about the Paradise fire. Paradise was a community that had a large number of retirees. It was affordable . . . These are people who retired, they're on fixed incomes, who lost everything. So, when you lose your home, and you don't have a lot of economic resources for rebuilding, you really have secondary emotional impacts . . . We need to make changes in our health care delivery as we confront the vast kinds of troubles that people are going to experience from climate change.[16]

How does it happen that older people migrate to destinations of disaster, whether by fire or flood? How is it possible that public officials are subject to the same denial in the face of reality? The Republican Congressman who represented the district including Paradise, California, rejected any link between wildfires and global warming, preferring debunked alternative explanations such as sunspots. We want any alternative except the reality that we ourselves are the cause of the problem—a point discussed later in the chapter "Who's to Blame for Global Warming?"

Perhaps we can best understand the meaning of the Dixie Fire or the Paradise Fire not from climate scientists or policymakers but from poets. Percy Bysshe Shelley (1792–1822) wrote that poets are the unacknowledged legislators of the world. One contemporary poet, Molly Fisk, born in 1955, is an elder who served as Laureate Fellow of the Academy of American Poets. She is now the poet laureate of Nevada County, California, near Lake Tahoe, not far from Paradise, California, where she encountered the fire herself. In this poem, Molly Fisk[17] expresses the catastrophe of the Paradise fire and the wreckage it left behind:

Particulate Matter

If all you counted were tires on the cars left in driveways and stranded beside the roads. Melted dashboards and tail lights, oil pans, window glass, seat belt clasps. The propane tanks in everyone's yards, though we didn't hear them explode . . .

Not to mention the valley oaks, the ponderosas, all the wild hearts and all the tame, their bark and leaves and hooves and hair and bones, their final cries, and our neighbors: so many particular, precious, irreplaceable lives that despite ourselves we're inhaling.

Molly Fisk

Molly Fisk's words give us images of our everyday world, like damaged cars on the road, as well as wild nature destroyed by fire and even the atmosphere itself that we're inhaling from wildfire smoke. We all are apt to think that disasters happen to someone else, in some other place, not really a threat to us. It's a natural defense mechanism, and I am subject to it, like everyone else. In the fall of 2020, I moved to California from a house in an open space, near Wonderland Lake, in Boulder, Colorado. Just a couple of months after we left, a massive fire broke out, the Marshall Fire, near Boulder, on December 30, 2020. It started as a grass fire, soon killing several people, and then, within days, it destroyed over a thousand homes with estimated damage of over half a billion dollars.[18] Before long 37,000 residents around Boulder would be evacuated, including friends and people I knew personally. Others fared much worse, losing their homes and a lifetime's possessions. Victims included people who lived far from where the fire first started. Within a year homeowner insurance premiums would rise an average of 17% throughout the state.[19]

For me, personally, the Marshall Fire was a huge shock. But should it have been? The truth is that in our beautiful house we had been living directly on the urban–wildland interface—an inspiring view of the Rocky Mountains but a threat from wildfire for which I had been almost entirely unconscious. True, my neighbor warned me about an earlier evacuation. But we all want to think it can never happen to me. The Marshall Fire was a shock to everyone: "In just a few hours—in some cases, minutes—fires engulfed entire

Figure 2.3 Marshall Fire (December 31, 2021) near Boulder, Colorado. Getty Images (milehightraveler).

homes and then whole neighborhoods. Families had little time to collect their belongings, if they could get back home at all."[20] I had been given a lesson in real time about the repeated message of this book: HERE NOW YOU. Like my granddaughter who saw the sky darkened with smoke at midday and wondered if she was going to school at night, here are lessons that both young and old will have to learn:

"Those climate tragedies are the direct result of how we plan and build communities. Federal, state, and local governments keep expanding highways, extending water lines, and building new infrastructure to meet the short-term demands of suburban growth. These infrastructure expansions tend to rely on traditional designs and technologies that pave over land, deplete natural resources, and discharge more pollution."[21] Americans in risky places will have to rethink what aging in place may mean for them.[22]

The real estate group Realtor.com reports that seven out of ten current home buyers have considered natural disaster risk in deciding where they will live.[23] What happened in Colorado was something much scarier than a wildfire[24] Wildfires are high on the list of concerns, and it's now possible to get information about fire risk with a tool named Fire Factor, which ranks properties according to their risk of wildfire damage in the coming decades. Here is what the future looks like.[25]

Those who are older may have heard about fires, including grass fires, but our defense mechanisms avoid the threat by asking, How bad could it be? The answer is not reassuring:

When wildfires spread through parts of Northern California wine country in 2017, they melted electronics, combusted cars and exploded propane tanks. The fires sent acrid smoke billowing into the sky, its footprint wafting over the state and extending for 500 miles into the Pacific Ocean.[26]

Consider the following dream by a writer, Madeline Ostrander, who visited Seattle when that city was afflicted by wildfire:

Trapped and Gasping

On Monday night, the smoke also descended on Seattle. I knew it was coming and closed all the windows, thought it was stuffy and warm indoors, and switched on a tiny air purifier that was never designed to clean my entire house.

On the evening of Labor Day, I dreamed that I was trapped in a campground with fire on all sides. My husband ran into the grass to find a way out but never returned. I tried to drive out but there was nowhere to go. I hunkered in a car, as flames rolled over it. Even in a dream state, I could feel heat. And then I woke, gasping.

The next day the sky was the color of cigarette stains, faintly luminous like a dying fluorescent bulb . . . My head was clogged with thoughts of burned towns, burned houses, burned lives. I watched the news. At night, as I tried to cook dinner, I burst into tears. All I could think about were the tens of thousands of people who were losing everything—and how onerous and terrifying it was.[27]

Madeline Ostrander's book is about what it means to be at home on an unruly planet. Her book consists of words and reports about facts about the world. But in her dream she encountered climate change differently: outside conventional time and space, in the inner "space" of the dream world. Then, after she woke up "gasping," she encountered the everyday world. But it was no longer "everyday." It was now "like a dying fluorescent bulb"—a powerful simile and a message about what it means to be trapped.

The dream leaves a challenge for the author and a challenge for us all. It was when Ostrander was trying to cook dinner that the feelings of the dream world intersected with waking consciousness: "I burst into tears." Isn't this the question that we ought to have, we who are climate advocates? We need to hear the deeper message: there's no escape from wildfire smoke.[28] Many Americans have seen this first hand when smoke from wildfires on the Pacific Coast was blown beyond and reached the Midwest and the East Coast.

How do we retain and encompass the genuine "the horror . . . the horror" of the situation, as conveyed in Madeline Ostrander's terrifying dream? These words, "the horror . . . the horror" were the last words of the "Apocalypse Now," a film about the Vietnam War. The words in the film are also the last words of Joseph Conrad's landmark novel *Heart of Darkness*. Conrad's novel tells the story of the leading character who has gone into the jungle searching for Kurtz, a rogue military leader who had gone over to the enemy and gone over into the very heart of darkness. The novel itself ends with Kurtz's own realization, "The horror! The horror!" as if there is nothing more to say.

Climate advocacy must remind us about the horror, but we must do more than that. We must hold out images of the good life that is possible, even when extreme weather events become more common. This is the message of the book *Deep Adaptation*.[29] In that perspective, the best approach to navigating climate chaos is to think in terms of local resources: farmers markets, community collaboration, mutual aid, and all the tools that help us recover agency and power over our lives. No one has stressed this message and demonstrated it better in his own life than the Eco-Elder Wendell Berry.

2.2   Wendell Berry: Eco-Elder

Wendell Berry, born in 1934, is both a farmer and an author. He lives an existence of "voluntary simplicity" and is widely known for his 1977 bestselling book *The Unsettling of America: Culture and Agriculture*. His warnings about the danger of consumerism offer compelling lessons for becoming a climate-conscious citizen.

Berry has an answer to the question of climate change "What Should We Do?" He believes that conventional responses to the problem require a moral solution: in his words, "the origin of climate change is human laziness."

Wendell Berry's life is an illustration of former House Speaker Tip O'Neill's quip that "all politics is local." Berry comes from a family of seven generations of farmers. He has lived on Lanes Landing Farm in Port Royal, Kentucky, for over half a century, abandoning an option for an academic career in New York. The environmentalist motto "Think globally act locally" illustrates his life since Berry is the author of over 40 books acclaimed around the world. In 2011, Barack Obama awarded him the National Humanities Medal.

At a national Slow Money Conference in Boulder, Colorado, I had a chance to hear Berry's daughter, Mary, present her father's ideas in person.[30] I became convinced not only to shop at farmers' markets but also to become a "slow money" investor in sustainable agriculture. Moving toward a plant-based diet for all of us needs to be matched by a better understanding of the ecological cost of the entire supply chain that brings food to our table.

Notes

1 See "Aging in Place: Growing Older at Home" at: www.nia.nih.gov/health/aging-place-growing-older-home See also: "Global Aging Is a Local Issue" at: www.publichealth.columbia.edu/public-health-now/news/global-aging-local-issue

2 Axios, "Climate is a Local Issue Now" at: https://growing-greener.org/ See also: "Global Aging Is a Local Issue" at: www.publichealth.columbia.edu/public-health-now/news/global-aging-local-issue

3 Dinah Pulver, "Climate Change Is Bad for Everyone. But This Is Where It's Expected to Be Worst in the US" *USA Today* (May 7, 2023) at: www.usatoday.com/story/news/nation/2023/05/07/what-are-the-worst-cities-and-states-for-climate-change-effects/11752271002/ The Urban-rural interface is critical here: "Climate Change, New Construction Mean More Ruinous Fires" (Jan. 22, 2022) at: https://apnews.com/article/climate-wildfires-science-environment-environment-and-nature-8d111d6f6dfa9bfa78a2fe9659802826

4 Christen Jaynes, "Smoke from California Wildfires Dimmed Solar Energy in 2020" *EcoWatch* (Dec. 9, 2022).

5 Quote from *Ecowatch*, 2022. See also Brianna Sacks, "California's 2020 Smoke Storm Was Horrific. What Did the State Learn?" *Washington Post* (June 10, 2023) at: www.washingtonpost.com/climate-environment/2023/06/10/smoke-pollution-crisis-california/ "Colorado Winter Wildfire Shatters the Illusion of Safety in Western Suburbs Los Angeles Times" Editorial, *Los Angeles Times* at: www.yahoo.com/news/editorial-colorado-winter-wildfire-shatters-110020185.html

6 Christen Jaynes, "'Climate Change Is Here': Every Part of the U.S. Will Suffer Climate-Related Disasters, Report Finds" *EcoWatch* (Nov. 17, 2023) at: www.ecowatch.com/national-climate-assessment-us-2023.html The results were far from local, reversing years of progress in U.S. air quality: Anita Hofschneider, "Study: Wildfire Smoke Is Reversing Years of US Air Quality Progress" *Grist* (Sept. 20, 2023) at: https://grist.org/wildfires/study-wildfire-smoke-is-reversing-years-of-us-air-quality-progress

7 Nora Mishanec and Sam Whiting, "Youth Activists Descend on S.F. Homes of Feinstein and Pelosi to Demand Climate Corps" *San Francisco Chronicle* (June 14, 2021) at: www.sfchronicle.com/local/article/Youth-activists-descend-on-S-F-homes-of-16247893.php

8 See: "Jane Fonda Reveals Why She Refuses to Retire at 85 and It's the Pro-Aging Positivity We All Need to Hear" at: www.womanandhome.com/life/news-entertainment/jane-fonda-reveals-why-she-refuses-to-retire-at-85-and-its-the-pro-aging-positivity-we-all-need-to-hear/

9 "Jane Fonda on the Climate Fight" *The Guardian* (Oct. 23, 2021) at: www.the-guardian.com/environment/2021/oct/23/jane-fonda-on-the-climate-fight-cure-for-despair-action

10 Alastair Gee and Dani Anguiano, *Fire in Paradise: An American Tragedy* (Norton, 2020), pp. 23–24.

11 ABC News, "Paradise Lost: Wildfire Chases Seniors from Retirement Havens to Field Hospitals" at: https://abcnews.go.com/amp/Health/paradise-lost-wildfire-chases-seniors-retirement-havens-field/story?id=59323835

12 Earth.org, "15 Largest Wildfires in US History" (Sept. 15, 2022) at: https://earth.org/worst-wildfires-in-us-history/?mc_cid=b5a0cdace7&mc_eid=50625bb884

13 Ibid.

14 Dani Anguiano and Alastair Gee, *Fire in Paradise: An American Tragedy* (W.W. Norton, 2021), p. 24.

15 Susan Garland, "'Do You Really Want to Rebuild at 80?' Rethinking Where to Retire" *NY Times* (Nov. 18, 2022) at: www.nytimes.com/2022/11/18/business/where-to-retire-climate-change.html?smid=nytcore-ios-share&referringSource=articleShare

16 "Climate Change Can Harm Mental Health of Older Adults" Q&A with Dr. Robin Cooper, by Ambika Kandasamy, *San Francisco Public Press* (June 16, 2023), cited in *Generations Beat Online*, ed. Paul Kleyman (Aug. 3, 2023). Dr. Robin Cooper is co-founder and president of the Climate Psychiatry Alliance.

17 Molly Fisk, "Particulate Matter" at: https://poets.org/poem/particulate-matter

18 Joseph Kane and Adie Tomer, "Financial Markets Fueled Colorado's Fires" at: www.brookings.edu/blog/the-avenue/2022/02/04/financial-markets-fueled-colorados-fires/ (Originally published in the Denver Post).

19 Benjamin Keys, "Your Homeowners' Insurance Bill Is the Canary in the Climate Coal Mine" *N.Y. Times* (May 7, 2023) at: www.nytimes.com/2023/05/07/opinion/climate-change-homeowners-insurance-housing-market.html

20 "'Stripped of Everything,' Survivors of Colorado's Most Destructive Fire Face Slow Recoveries and a Growing Climate Threat" at: https://apple.news/AdH4wCvv7QPCLx8OTuQQrNg

21 Kane and Tomer, "Financial Markets Fueled Colorado's Fires."

22 "Why Does the American West Have So Many Wildfires?" at: www.nytimes.com/2022/08/01/climate/wildfire-risk-california-west.html?referringSource=articleShare "Rethinking Risk and Responsibility in the Western Wildfire Crisis" at: https://ssir.org/articles/entry/rethinking_risk_and_responsibility_in_the_western_wildfire_crisis?utm_source=Enews&utm_medium=Email&utm_campaign=SSIR_Now#

23 "70% of Recent Home Buyers Considered Disaster Risk in Deciding Where to Live" at: https://yaleclimateconnections.org/2022/09/70-percent-of-recent-home-buyers-considered-disaster-risk-in-deciding-where-to-live/?utm_source=Weekly+News+from+Yale+Climate+Connections&utm_campaign=f7fb3f56ac-EMAIL_CAMPAIGN_2022_09_15_03_46&utm_medium=email&utm_term=0_e007cd04ee-f7fb3f56ac-59379135

24 "The Return of the Urban Firestorm" (Jan. 22, 2022) at: https://nymag.com/intelligencer/2022/01/colorado-saw-the-return-of-the-urban-firestorm.html?utm_source=pocket-newtab

25 "Risk of Very Large Fires Could Increase Sixfold by Mid-Century in the US" at: www.climate.gov/news-features/featured-images/risk-very-large-fires-could-increase-sixfold-mid-century-us

26 *Inside Climate News* (Apr. 23, 2023) at: https://insideclimatenews.org/news/23042023/wildfire-smoke-pregnancy-children/

27 Madeline Ostrander, *At Home on an Unruly Planet: Finding Refuge on a Changed Earth* (Henry Holt, 2022), p. 163.

28 David Wallace-Wells, "There's No Escape from Wildfire Smoke" *NY Times* (May 17, 2023).

29 Jem Bendell and Rupert Read (eds.), *Deep Adaptation: Navigating the Realities of Climate Chaos* (Polity, 2021).

30 "Slow Money's 2013 National Gathering in Boulder, Colorado, April 29–30 Brings Juice to Local Economies by Connecting Food Businesses with Funding" *Cision* (Mar. 14, 2023) at: www.prnewswire.com/news-releases/slow-moneys-2013-national-gathering-in-boulder-colorado-april-29–30-brings-juice-to-local-economies-by-connecting-food-businesses-with-funding-197997761.html

3 Flood

The Water Will Come[1]

The sea, the great unifier, is man's only hope. Now, as never before, the old phrase has a literal meaning: We are all in the same boat.

—Jacques Cousteau

Sea levels are rising constantly, every single day, without exception. This means that coastal areas around the world are vulnerable to storm surges intensified by greenhouse gases. Flooding is a special threat to coastal areas of the USA. Coastal counties are home to over 128 million people or almost 40% of the nation's total population. Twenty-six states—more than half—have ocean, sea, gulf, bay, or lake coastlines.

That geography means that millions of people are vulnerable to coastal flooding. Some people—like many residents in Hilton Head Island, where my wife and I used to own a time-share property—have put their houses up on stilts. It's one version of adaptation. But it isn't complete protection. Future economic vulnerability is also serious since ocean activities make up more than 55% of the gross domestic product (GDP) in coastal areas, including ocean transport as well as recreation, tourism, seafood, and resource extraction. Rising seas are a challenge for coastal cities like New York, Miami, and New Orleans, and many others as well.[2]

Climate change and sea level rise present a clear threat to older homeowners who want to age in place.[3] Coastal populations will get older—and more fragile—as young people flee rising seas, according to new academic research[4]

With rising sea levels and increasing storms, 40 million Americans are living in areas prone to flood. For understandable reasons, we know that millions are older people who want to age in place. The nonprofit research group Climate Central has estimated that hundreds of thousands of properties in the USA could find themselves submerged by tidal lines in the coming years.[5] As sea level rises, there will be a shift in coastlines and property lines. The magnitude is staggering: more than 4 million acres of land could fall beneath new tidal boundaries within the next 30 years. Thirty years is the time frame for many home mortgages today.

DOI: 10.4324/9781003345992-4

Jacques Cousteau has said that, when it comes to flooding, we're all in the same boat. Threatened, yes, but not all will be sinking. Different people will be affected in different ways depending on geography. Different states and regions will be affected in different ways. Homeowners on the Gulf Coast or the Atlantic Ocean are at greatest risk. For example, more than 8% of the entire land area of Louisiana could disappear. Other states including Florida, Texas, and North Carolina will also be under flood threat.

For those who aim to age in place, sea level rise means that the water will come to their door. They may find themselves aging in the wrong place.[6] For example, homeowner insurance in 2023 was expected to increase by 40%.[7] Americans of all ages are abandoning their dwelling place because of flood risk and climate change. Can elders migrate, too?[8]

For older homeowners who have paid off their mortgage, it means that home equity—often the largest part of their wealth—is at risk. Those with mortgages are likely to face increased property insurance costs. Even if they give up on hopes for aging in place and want to relocate, they will face diminished value of properties threatened by sea level rise. Aging in place in a time of climate risk has its price. Roy Wright, previously in charge of insurance at the Federal Emergency Management Agency (FEMA), said it best: "Risk has a price. We're just now seeing it."[9]

Climate change for elders is more than just a matter of public policy:

Climate change is personal. It is not abstract. The warming climate impacts our economies, influences our politics and culture, threatens the food we eat and the water we drink; it even affects our love lives. As climate change accelerates and extreme heat and climate disasters displace more people around the world, the crisis is increasingly disrupting our fundamental sense of where we belong and what we consider home.[10]

Arthur C. Clark said, "How inappropriate to call this planet Earth when it is quite clearly Ocean." When we look at rising sea levels we recognize the ocean as a threat, which is certainly true. But there is much more to recognize, as oceanographer Sylvia Earle has said:

The greatest diversity and abundance of life is out beyond where most people ever go. You don't see your heart either, but you know it matters. You don't see the ocean with all the life that's in it . . . It is easy to find despair, but don't let it own you. Don't let the bad guys win; be a force for good.[11]

Sylvia Earle also said: "Far and away, the greatest threat to the ocean, and thus to ourselves, is ignorance. But we can do something about that" (*EcoWatch*, July 27, 2023).

3.1   Sylvia Earle: Eco-Elder

Sylvia Earle, an advocate for ocean conservation, is the first woman to lead the National Oceanographic and Atmospheric Administration. She holds the record for the deepest walk on the sea floor and is a world-renowned expert on marine biology. Now in her late 80s, Earle has witnessed disruptions to sea life from pollution, overfishing, coral decline, ocean acidification, and, above all, climate change.

The ocean is important: more than two-thirds of the surface of the earth is covered by the sea, and that's where 97% of our planet's water is found. Yet 95% of the ocean remains still unexplored. For climate change, the oceans are absolutely important because they absorb a quarter of the carbon dioxide we're putting out. Ocean acidification is one consequence, with an adverse impact on life in the ocean.

Sylvia Earle has called the ocean the "lungs of our planet" and, because the sea is under serious threat, she said, "So are we." Her book, *National Geographic Ocean: A Global Odyssey*, is intended to sound the alarm because "now is the best chance we'll ever have to protect the ocean and to take action." Like David Attenborough, Earle says we are seeing a diminished planet, which she has witnessed in her lifetime. And, like Attenborough, she insists that it's not too late since we have the knowledge to reverse the decline.

This makes the time in which we are living a crucial moment: "We are at the most exciting time maybe ever to be a human, because we're armed with knowledge."

An Eco-Elder, Sylvia Earle has devoted the last years of her life to protecting the ocean and the planet for future generations.

Threats from floods and sea level rise are not hypothetical. On the contrary, the threat is already "baked in" so that even if we end all carbon emissions tomorrow, even if we turn off every single power source on the earth, the water will still be coming for decades to come. The bottom line is clear: the water will come, just in the same period when the American population is growing older. Maps will show us how rising water will affect each of us.[12] This is our condition: HERE NOW YOU.

The six strongest hurricanes on record have all happened since 2017, so, as the saying from the Apollo 13 space flight goes, "Houston, we have a problem." In the case of hurricanes and extreme weather events, the problem is not limited to Houston or Texas. When 10,000 Boomers turn 65 each day, the problem is here and now for those who want to age in place. For example, when Hurricane Sandy hit New York in 2012, almost half of those who died were over age 65.[13]

A case of what aging in place could be like in years to come is Oakwood Beach on Staten Island, New York, one of the five boroughs of New York City. The town is less than an hour's drive away from Manhattan, where I used to live. Once a thriving working-class neighborhood, Oakwood Beach was devastated by Hurricane Sandy. In the aftermath, the New York State government bought out most homes and it now remains largely empty. But some older people stayed behind. This is what aging in place looks like.

For example, Lois Kelly, age 71, lived in Oakwood Beach, long before Hurricane Sandy. "It was a lovely neighborhood, like a little community," she said. When Sandy hit, she and her husband were trapped in their one-story house as it filled with water. They got enough money from the insurance company to rebuild and chose not to relocate. Oakwood was their home, and they didn't feel like starting over.

Kelly's situation reminded me of my own hometown, Merrick, on the south shore of Long Island, which was also hit by Hurricane Sandy, as it was by hurricanes in my childhood in the 1950s. But Merrick recovered and did not become a ghost town like Oakwood Beach.[14] Of course, residents of my hometown of Merrick now face higher flood insurance rates.[15] As for Oakwood, estimates from the National Oceanic and Atmospheric Administration warn that it could become entirely flooded in only three decades. Instead of rebuilding and aging in place, there is an alternative of managed retreat—another form of adaptation to climate change in an aging society.

However we respond, the ocean floods will have an impact on our minds as well as on our homes. The terror comes out in our dreams at night. Here is a dream from a resident in Alaska, who faced the impact of flooding.

Rising Waters

In the dream, a storm came and Betsy Bekoalok watched the river rise on one side of the village and the ocean on the other, the water swallowing up the brightly colored houses, the fishing boats and the four-wheelers, the school and the clinic.

She dived into the floodwaters, frantically searching for her son. Bodies drifted past her in the half-darkness. When she finally found the boy, he, too, was lifeless. "I picked him up and brought him back from the ocean's bottom."[16]

Our dreams convey our deepest fears, but they can also convey intimations of hope.

A Bridge

I dreamed I went on a field trip to a nature reserve set up for endangered species. We walked by an apple orchard that wasn't finished yet and someone in the group commented that it was for the zebras but as long as the people watching the zebras did a good job they wouldn't need to relocate them to the orchid. A bridge to the secret lake that appears in many of my dreams was washed out because of the raised sea levels.[17]

Figure 3.1 View of New Orleans after Hurricane Katrina. Public domain.

The reality of endangered species is what we now call the biodiversity crisis, and the dream "A Bridge" begins with Noah's ark symbolism. The dream suggests an aspiration toward hope, a recurrent image for this dreamer conveyed in the phrase "the secret lake that appears in many of my dreams." This dream contains both hope and fear: the bridge itself could be washed out because of rising sea levels.[18] But, in reality, not just in dreams, the water will come.

Case Study: Hurricane Katrina in New Orleans

Devastating hurricanes are part of American history. The hurricane that struck Galveston in 1900 was America's deadliest natural disaster, killing more than 8,000 people, well before a time when global warming would have intensified the impact. In 2005 Hurricane Katrina in New Orleans killed more than 1,800 people, disrupting thousands of lives and damaging or destroying more than a quarter of a million homes.[19] Two-thirds of those who died from Katrina were over the age of 65, so the storm represents a major example of an extreme weather event for an aging society.[20] The death rate of elders is also related to a preference for aging in place. Along with others, older residents believed they could handle the threatening storm. One poll found that a clear majority of those who did not evacuate misjudged the severity of the storm that was coming: they thought it wouldn't be as bad as predicted, or "I survived Hurricane Camille" or sentiments along these lines.

I heard some of these stories myself when I went to New Orleans a few months after Hurricane Katrina hit. I was invited at the request of the Home Care Agencies of Louisiana, as they tried to understand the terrible toll inflicted on homebound elders. The day I arrived at the New Orleans Airport, I immediately recognized that, even after a couple of months, things were not back to normal. Restaurants and shops in the Airport were mostly dark and closed, as if in a ghost town. During my time with the home care agencies, we had a chance to take a tour of some of the devastated areas of New Orleans. It reminded me of pictures my father had sent of Germany in the early 1950s when I was a small child and he was called up and stationed in Germany, which still bore the scars of WWII.

The Home Care Agencies confirmed that older residents repeatedly failed to evacuate until it was too late to do so. The stories in the press about how elders were abandoned in nursing homes or hospitals completely failed to capture the problem: aging in place was a choice that resulted in death. "I'm not going without my pet" "It's too risky." Some home care agencies were successful in convincing older people to leave: for example, by engaging family members or caregivers, having a distant daughter make a plea, "Do it for me, don't stay."

As a biomedical ethicist,[21] I quickly saw that the death sentence that fell on so many older people was the result of a fatally limited idea about individual autonomy. Does informed consent and the right to refuse treatment mean just listening when people say no? Why can't it include engaging people in the family when refusal of services means death? It turned out that creative home care workers were able to save the lives of elders at risk because when people refused to evacuate, the workers consulted with family members, some at a distance, as in the daughter-in-law who said, "Do it for me. Leave your place, just in case." That's how some lives were saved.

One result of my consultation with the Home Care Agencies was developing a systematic protocol for evacuation under threats like what we had with Hurricane Katrina. The Louisiana Home Care Agencies knew perfectly well that these threats would not go away in the future. Some who used creative casework could put those lessons to use in the future and save lives.

What I saw in New Orleans in 2005 was only the outer face of destruction. The hidden story is that the impact of damage depended very much on the socio-economic status and inequality of people in New Orleans who experienced the hurricane. Naomi Klein put it succinctly, "It's inequality that kills."[22]

Mary Heglar,[23] writing in "After the Storm," tells the story of what happened to New Orleans and especially its impact on African-American elders. As she describes it, electricity was cut off and later came back on:

When the power came back on, we saw everything with our own eyes.
We saw that the town on the coast had been completely washed away.

I can still hear Governor Barbour's voice: "I don't mean they were badly damaged. I mean they're simply not there."

As a Black woman from the South, she writes:

Thanks to Katrina, I can't look at the climate crisis without seeing the grimy fingerprints of slavery and Jim Crow.
  I myself saw the devastation from Katrina a couple months after it happened, and it reminded me of pictures my father had sent me of Germany when he was stationed as a soldier in the years after W.W.II. When Mary Heglar visited New Orleans ten years later The grit of the storm was still there, on every billboard, every building, every face . . . To this day, everything there is dated as either "before the storm" or "after the storm"—and no one questions which storm.

Hurricane Katrina had an impact on the young and the old, people of many generations. For example, Varshini Prakash, born in 1993, is co-founder of the Sunrise Movement, the youth-led group fighting to stop climate change. She co-edited *Winning the Green New Deal: Why We Must, How We Can*. Here's how Hurricane Katrina affected her:

We [young people] always kind of had this fear of this looming crisis. One of the experiences that defined my childhood was hearing about Hurricane Katrina. I was 12. You know, seeing these images of people on their roof, hearing about bodies just floating downstream. And the government doing nothing to support those communities.[24]

Two-thirds of the victims of Hurricane Katrina were 60 or older, and that was comparable to the high death rate for elders in the Chicago heat wave of 1995. In other words, these "extra deaths" were primarily older people. As in Chicago, the mortality was concentrated in poorer, Blacker neighborhoods. Margaret Morganroth Gullette, author of *Ending Ageism, or How Not to Shoot Old People* and *Aged by Culture*, has said that what happened in Hurricane Katrina is a case of Eldercide.[25] She points out that the losses were untimely, premature, and preventable—therefore not "natural" in the way say speak of a "natural disaster." Gullette argues that the lesson taken from Katrina was city-wide "climate-disaster preparedness," instead of the need to resist multiple forms of bias, including ageism. When we think about climate change in an aging society, we have to look beyond prevention or adaptation and look at deeper questions about climate justice.
  The threat of flood is not limited to the USA. Antonio Guterres, Secretary General of the United Nations, has warned that, on a global basis, sea level rise could drive one out of ten people from their homes.[26] Flooding is clearly a serious threat demanding a global response.[27] When we're talking about 900

million people at risk, it means that sea level rise could mean a mass exodus on a biblical scale, a point we will return to in pondering migration as one response to climate risk and aging in place.[28]

When we speak about threats from floods, including rising seas, the impact varies by geography. For example, here are eight American cities that could vanish by 2100 because of flooding:[29]

Atlantic City, New Jersey
Charleston, South Carolina
Houston, Texas
Miami Beach, Florida
New Orleans, Louisiana
Boston, Massachusetts
New York City, New York
Virginia Beach, Virginia

Two of these cities—New Orleans and Miami—could disappear much sooner: by 2030, according to some estimates.[30] We will examine both cities in more detail further on. But, beyond that, we're also talking about cities like New York City, Philadelphia, Boston, Seattle, and Jacksonville, Florida.

Case Study: Philadelphia

It would be a mistake to think of climate risk and aging in place as a threat limited to California or the Southeastern USA. Hurricane Ida in 2021 gave the city of Philadelphia lessons that helped mobilize action in anticipation of future hurricanes:[31]

> [Hurricane] Ida swept through the Northeastern U.S. and Philadelphia, bringing torrents of rain, destructive tornadoes and high-speed winds. In Pennsylvania, Ida's remnants caused millions of dollars in property damage, swamped the Vine Street expressway, overwhelmed water plants, destroyed homes, and killed four people. More than 83,000 households in Pennsylvania applied for FEMA assistance in the aftermath of the hurricane.

Global warming makes hurricanes more intense. Scientists say that storms like Ida, once atypical, will become the new "norm," and Ida is just one example of the extreme weather that cities now face. Depending on where you are, the present—and the future—looks like blazing heat waves, devastating wildfires, and regular flooding, or some combination of the three.

Climate adaptation became a major theme for the COP27, the annual U.N. climate conference, in November 2022 in Egypt. A global conference commands our attention. But the slogan for environmental action reminds us of something else: "Think globally, act locally."

Philadelphia, on the East Coast of the USA and in the path of future hurricanes as much as Miami, is grappling with a basic challenge:[32]

Philadelphia shows both the promise of cities to be leaders in adapting to a rapidly shifting climate—as well as their limitations. The questions that confront Philadelphia are echoed in other urban environments: what kind of policies, technologies and strategies will be needed to weather the rollercoaster of climate consequences in a given place, without leaving the most vulnerable residents behind? What makes any single location worth preserving or abandoning, and who gets to decide whether to stay or to go? What kind of regional, national and international aid will cities need to survive, and what will that survival look like?

. . . Do you want to invest in people or do you want to invest in places? Until we take a hard look at that question and its implications . . . we're not really having a fully coherent conversation around cities adapting to climate change.

Sometimes those goals overlap; cities' population density makes them ideal for things like cooling stations, for example, because they can serve larger numbers of people than they would in rural areas. But the goals of saving people and of saving a specific place don't always align. "It's a hard decision to stay attached, and it's obviously a hard decision to give up that attachment," Hughes said. "Do you want to make a wet place dry? Or do you want to move to a dry place?"

Hurricane Season Everywhere

In late October 2012, Hurricane Sandy hit New York City with a devastating impact. Water poured into lower Manhattan subway lines, and 2 million Con Ed customers lost power. Older people were among those most severely affected. In Hurricane Sandy, as in Hurricane Irene one year earlier, older people with limited mobility or in long-term care settings were vulnerable to an extreme weather event. The International Longevity Center, located in New York City, documented something similar for older city residents after September 11, 2001. People living alone are more vulnerable to social isolation in times of crisis and stress. Tens of thousands of older people living alone during Hurricane Sandy had disabilities making evacuation response very difficult.[33]

There are many causes of flooding, but the overriding issue is rising sea levels, caused by climate change and melting Arctic and Antarctic ice.[34] Climate change may be the future we don't want. But even if we shut off all greenhouse gas emissions today, the momentum built into the system will cause sea levels to rise for decades to come.

There are serious economic implications here. Rising seas from climate change could threaten $34 billion in real estate just in the coming decades.[35] Rising sea levels put Miami and New York at big risk for flooding of commercial properties:[36]

About a tenth of the world's residential property by value is under threat from global warming. The severe weather brought about by greenhouse-gas emissions is shaking the foundations of the world's most important asset class. The potential costs to homeowners are enormous . . . Homeowners face a $ 25 trillion bill from climate change: Property, the world's biggest asset class, is also its most vulnerable . . .

The severe weather brought about by greenhouse-gas emissions is shaking the foundations of the world's most important asset class. It is a huge bill hanging over people's lives and the global financial system. And it looks destined to trigger an almighty fight over who should pay up.[37]

These trends point to a major policy problem for climate change in an aging society:

Local emergency managers know all too well which places in their communities should not be built back after a storm. But they are rebuilt, because the federal government and states provide multiple incentives to rebuild rather than to relocate. The assumption is that taxpayers will always be there to back up private investment after even predictable natural hazards.[38]

Many elders are likely to be left behind when others, more mobile or more wealthy, can leave. But being left behind is not pleasant.[39] In terms of life experience and psychological impact, it can be a trauma like death.[40] Aging in place can mean literal death, which is what happened in Hurricane Ian in Florida in 2022.[41] Hurricane Ian was a Category 4 storm, one of the costliest in Florida history, with storm surges as high as 15 feet and more than 20 inches of rain in some places.[42]

The sheer magnitude of Hurricane Ian can be appreciated by comparing its impact with comparable disasters around the world:

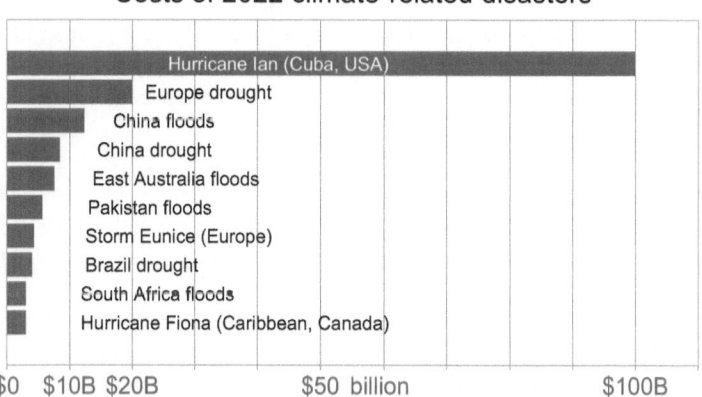

Figure 3.2 Costs of 2022 climate-related disasters. Public domain.

What was the impact of the hurricane on elders aging in place? Preexisting chronic conditions were a factor, but aging in place meant that nearly 60% of those who died were 65 or older.[43] "There is no one who is required to make sure they evacuate or that their home environment will keep them safe," said Lindsay Peterson, an assistant professor who conducts disaster preparedness research at the University of South Florida's School of Aging Studies. "They are much more vulnerable, and we see that in these statistics." The reports suggest many would still be alive had they evacuated. Nine people died because power outages meant they could not operate oxygen or dialysis equipment, including a 70-year-old diabetic in Charlotte County who went a week without dialysis.

A similar pattern is evident elsewhere around the USA. In Louisiana, in 2019 residents of Isle de Jean Charles were among the first communities the federal government is working to resettle en masse in the face of rising seas. New Jersey has overseen buyouts of hundreds of properties since Superstorm Sandy in 2012. Harris County in Texas has undertaken more than 800 buyouts since Hurricane Harvey in 2017 and approved 1,600 more. Many smaller communities have followed a similar path, where elders become the first climate refugees, even if they won't be the last. Painful and difficult questions will persist.[44]

In Myrtle Beach, South Carolina, Keith and Tyra Moore are an example of a couple who planned to spend their retirement in what turned out to be a flood zone. After experiencing multiple floods, the couple is now planning for a home buyout and managed retreat. Keith Moore put it this way: "We used to live on the creek, then we were in the creek, and now we are up the creek." But their neighbors don't have any plans to leave.

We all console ourselves or deceive ourselves, with the thought that it won't happen to us. But then it does. On the very last day of 2022 (New Year's Eve), my wife and I unexpectedly ended up driving to our home in San Mateo through an atmospheric river that engulfed the San Francisco Bay Area, drenching us with the second-worst rainstorm in 170 years. In recent years we've begun to hear more about these "atmospheric rivers," downpours fueled by a band of moisture that can extend for a thousand miles.[45] Meteorologists tell us that an atmospheric river can carry two or three times the water in the Amazon River. Atmospheric rivers can be helpful: for example, in building up snowpack, which is more important than ever during periods of drought. In California, where I live, atmospheric rivers also intensify the risk of a megaflood, which has happened before: for example, in 1862. Climate change influences atmospheric rivers,[46] and the risk of flood is likely to be worse in years to come. Our experience with the New Year's Day flood was a glimpse of what could lie ahead.[47]

As we drove through streets and encountered floods, barely able to make it home, I wondered: Can all of this simply be a bad dream? Meteorologists tell us that unexpected flooding is likely to be more frequent in years to come. In retirement my wife and I live at the top of high hills on the Bay Area

Peninsula. Even if we were not actively searching for safety in old age, we thought we had found it. But we were wrong. It was not a bad dream, but a reality, the "new normal." Such intimations of flooding appear in this actual dream "The Flood Is Coming."

> I dreamt that the clouds were very dark, stretching out over my town, and the rain and wind were so bad we had to stay in the house. I looked outside at the clouds, and saw some big billowing clouds below us. I realized they were rolling down the street, and were not clouds at all, but a torrent of floodwater. I knew the water could come in through our front door, and I woke up from the dream in a panic of thinking about how to get onto our roof to be rescued.[48]

Our dreams show us what we already know but cannot yet see.

Halting the Flood

Hindu tradition often invokes the idea of "karma," which is a concept hard to translate easily. Karma is not punishment or destiny or anything we can quickly wrap our minds around. But when we think about the threat of flood and climate change, there is wisdom from the great spiritual traditions that point in the direction we need to go. Karma does not mean "we are doomed." But it does mean that our condition, our destined struggle, is not an accident, not something that fell upon us from high without our responsibility. On the contrary, as Clarissa Pinkola Estes says, "We were made for these times."[49] To act rightly in these times, we need wisdom or what the Tibetan Buddhists call "skillful means."

One of the sources for wisdom is the following folk tale,[50] drawn from Indigenous Australian Aborigine traditions:

"Halting the Flood"

In the Dreamtime, two brothers walked through a desert. The older brother carried a large water bag but refused to give his younger brother a drink despite many pleas. The older brother went off to go hunting, carefully hiding his water bag before leaving. The younger brother searched for the water bag, found it and hit it with a club. Water burst out from the bag, and soon threatened to cover the whole land. The older brother ran back and tried to close his bag, but to no avail. Soon the water formed a great ocean and the two brothers drowned.

When the Bird Women learned of the flood, they flew in from different parts of the land—the bluebird, parrot, bell bird, and all the other bird families. Each Bird Woman carried roots of the kurrajong tree. (Those roots store potable water and are essential for surviving in the desert.) The Bird Women placed the roots along the seashore, to absorb the water from the rising ocean and in this way stopped the flood. After saving the land, the Bird Women returned to their homes.

We speak of climate change today and note the rising level of oceans around the world, as well as extreme weather events that cause local flooding in many places far from the sea. We even consider these to be "natural disasters," but are they really "natural"? We recognize the role of humanity, expressed in a phrase like "anthropogenic climate change." The causal connection is evident.

The Australian Indigenous peoples expressed causality in the language of Dreamtime, but in the "Halting the Flood" story, humanity is also the agent there, too. In this folk tale the older brother, as in our industrialized societies of the Global North, must take responsibility for climate change. But, just as today, the older brother refuses to give those who are vulnerable the means of coping with the flooding that has already begun. Countries that are low lying, countries in the Global South, cannot halt the flood without help. Instead of "carefully hiding our water," we set up barriers to prevent migration for people whose livelihood has been torn from them by extreme weather events. In this tale the younger brother hides the water bag and unleashes disaster. At international COP climate conferences, countries at all levels come together to find ways to stop the disaster. But have they stopped hiding the water and unleashing disaster?

In this folk tale who are the Bird Women saving the land? It is not the male creator or trickster god and not even the water serpent that brings rain, as Allan Chinen points out. Could the Bird Women be the clan elders for each bird clan (bluebird, parrot, etc.)? Could these elders come to save us, or will they arrive too late? As in "Halting the Flood," we can try to close the bag but to perhaps no avail: "Soon the water formed a great ocean and the two brothers drowned."

How soon can someone come to save us from the flood? How much time do we have? 2030? 2040? In this tale the Bird Women do come to the rescue, each carrying roots, techniques for survival. The Bird Women place these roots along the seashore to absorb the rising ocean. Will we find such techniques to restore the melting glaciers of Greenland or Antarctica? Will we save the land for our successors? Will we become Good Ancestors? Or is "Halting the Flood" just a tale of wish fulfillment with a happy ending? We cannot know at this moment, which is why we must act in the best way we know how.

Insurance

Extreme weather events, both fire and flood, threaten more and more people and properties. The critical questions are now arising: Who will cover the losses? Who will pay to rebuild communities? Can private insurance, or public insurance, actually meet the needs that are now arising? These are the

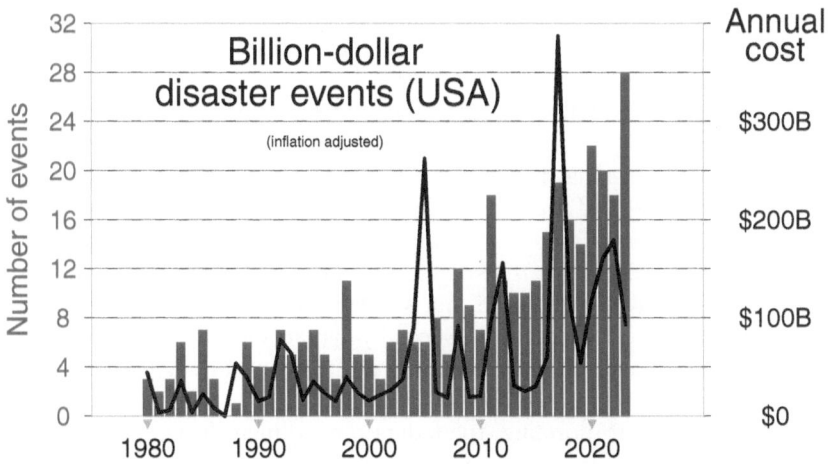

Figure 3.3 Billion-dollar disaster events (USA) based on NOAA's National Centers for Environmental Information. Public domain.[51]

questions we will have to face with climate change in an aging society. Natural disasters are now costing the U.S. insurance industry $100 billion a year. What happens if there's no one to pick up the tab?[52]

Climate risk comes down to insuring against the inevitable: the water will come; the fires will burn. Those who can afford to buy insurance may want to do so, but can the poor afford to pay for insurance against climate-related disasters? There is a profound unfairness when those who profited least from fossil fuels—the poorest—end up paying for the damage. We will come back to this question again in the next chapter when we ask, pointedly, Who's to blame for climate change?

Is insurance the answer to home vulnerability from climate-induced fire or flood? According to U.S. Census data, approximately two-thirds of Americans own their own homes and home equity accounts, on average, amount to about half of the financial net worth of those who are homeowners. A primary residence has been the largest single asset for all households across the life course. Older people obviously have more equity in their homes than younger people. But when older people have paid off more of their mortgages, they are more directly vulnerable to losses from fire or flood. With aging, it becomes more difficult to move, especially when friends or relatives live nearby and are important agents for caregiving.

Thus, the question becomes important: can insurance give us protection against fire or flood? In 2023 State Farm, America's largest insurance firm, stopped accepting new homeowner insurance applications from the state of California.[53] Robb Lanham said, "I never thought I would see in my lifetime houses that are flat-out uninsurable." Lanham is the chief sales officer for insurance brokerage HUB Private Client, covering up to 500 insurers. He noted that hard-hit states were not limited to California but included Florida, Texas, Colorado, Louisiana, and New York, all states where fire and flood from climate change have been a factor.[54]

Insurers are already fleeing from disaster-prone states:[55]

California. Louisiana. Florida. Who's next? The insurance markets in these hurricane- and fire-prone states have descended into turmoil over the past few years as private companies drop policyholders and flee local markets after expensive disasters. State regulators are stepping in to stop this downward spiral, but stable insurance markets will mean higher prices for homeowners, especially in places like low-lying Miami, where the average insurance premium is already around $300 a month.

When it comes to flooding, the federal government has created national flood insurance options[56] but climate change is likely to make public as well as private insurance options less secure. Climate change means increased insurance rates.[57]

Fire insurance is typically required to have a mortgage, and we can find out if our home is vulnerable to flood. But can we pay for insurance? And will we collect when fire or flood comes to our door? I found out it was not so easy when our two units in Boulder, Colorado, got flooded. Yes, you can collect from insurance companies, but companies usually don't make it easy to collect. And, as with travel insurance, sometimes you can't collect at all. A case in point is Hurricane Ian (September 2022), which was a warning to make flood insurance accessible to everyone.[58] But not everyone can pay for it. Climate change is making some homes too costly to insure.[59] Even worse, what happens when insurance companies cut their payouts far below the damage caused by the hurricane?[60] The insurers did this by actively trying to limit payouts approved by licensed adjusters who were trained to assess home damage. The result was that many homeowners were left to pay the bill for repairs from flooding from this Category 4 storm. As we've seen, many of the deaths from Hurricane Ian were elders.

What about the insurance companies themselves? It seems clear that climate change is hurting insurers, and they know it.[61] Some have argued that insurers could have been climate heroes—by bringing news and "reality therapy" to the public and promoting action to avoid climate change.[62] But instead they have risked a crisis that could dwarf the financial crisis of 2008, or else they simply leave a region, as many have left Florida because the risks are too great.

Risky Business: Underinsured Against Climate Disaster

In recent years, hundreds of thousands of people in high-risk disaster areas across the USA have been dropped from their insurance policies, leaving them both physically and financially vulnerable. At the same time, premiums have skyrocketed, making insuring homes and businesses out of reach for many. And federal insurance and relief programs have come under scrutiny for payouts that contribute to inequality.[63]

The insurance industry wasn't set up to account for climate change, which is increasing the frequency, scale, and severity of disaster claims. From flooding in Appalachia to fires in the Pacific Northwest to hurricanes wreaking havoc from Puerto Rico to Nova Scotia, we've seen frequent and fierce weather take lives and devastate communities.

As more people and property face loss due to extreme weather events, who will pay to protect and rebuild communities? And what policies are being constructed to help the insurance industry stay afloat?[64]

Adaptation for Fire and Flood?

Well, don't look for it in recent legislation on climate change. True, the 2022 Inflation Reduction Act authorizes $369 billion to respond to global warming concerns. But how much of that money would go to insuring that houses can withstand fire and flood? Actually, nothing at all.

We're not talking about mitigation or reducing the scale or the impact of climate change. Mitigation is clearly important, but adaptation has a place, too. The Inflation Reduction Act, whatever its virtues, doesn't provide money for resilience to climate disasters, even though all sources forecast an increase in natural hazards in years to come. This was the view of Gabriel Maser, an executive with the International Code Council, the nonprofit group responsible for model building codes.

Personal home equity accounts for nearly 30% of the net worth (total assets) of Americans, and home equity represents a huge portion of net worth for older Americans who've paid into mortgages for decades. The Federal Reserve Bank (St. Louis) estimates homeowner equity at $30 trillion. But net worth is at serious risk should flooding occur. Unfortunately, accounting for flood risk would diminish house prices by an estimated $187 billion. In contrast to other types of risk where insurance is possible, private insurance is generally not available for residential coverage for flooding.[65]

Who Gets Insurance

What about insurance? Across North America, only 40% of economic losses from household disasters are insured. Even for those living in a FEMA-designed 100-year flood plain, only 30% have flood insurance and only 40% of renters have renters' insurance.[66]

Homeowners, especially older homeowners who have aged in place, are likely to believe that "I'm covered by insurance." That's especially true if they haven't tried to make a claim lately. Yet the fact is that hundreds of thousands of people living in areas at high risk of disaster have been dropped from insurance cover or could be dropped when big flooding begins. When premiums rise dramatically, they may be unable to pay higher costs. The insurance industry sets premium rates based on past experience. When things change because of climate change, the insurance industry may not be prepared any more than homeowners are for the "new normal" of climate change.

Older homeowners, particularly those whose net worth is tied up in their homes, are at great risk for damage from fire or flood. The First Street Foundation is a nonprofit trying to improve understanding of climate-related risks. Along with the Environmental Defense Fund and Resources for the Future, they have calculated the estimated annual losses from floods over the next few decades. Researchers found a big discrepancy in value in counties along the coasts.[67] These are places where disclosure of flood risk is not required for real estate transactions, so homeowners may not be aware of the climate risk involved. It leaves unclear who will bear the financial risk and uncertain about how the housing market will respond.

Fire and flood are both risks from climate change. The average number of large U.S. wildfires has risen by 30% over the past 15 years and by nearly a fifth in just the last five, according to Lloyd's of London insurer Chaucer. "The tried and tested approach, where decades' worth of historical data serve to estimate future claims, falls short when weather patterns change and hurricanes, floods, heat waves or snowstorms become more extreme and unpredictable."[68]

During Hurricane Ian (2022), there were areas of maximum danger where residents were told to evacuate. But fewer than 19% of homes in those counties had coverage through the National Flood Insurance Program.[69] The insurance industry in Florida is already fragile, and the state government has stepped in to try to cover the gap.[70] There have been warnings that a major storm like Hurricane Ian (2022) could push some insurance companies into bankruptcy, with disastrous impact on people trying to collect on their claims. A significant number of insurance companies are already on the danger list compiled by state regulators in Florida. Some major companies, such as Allstate and State Farm, have already reduced their exposure in Florida over the past decade.

Is the solution for the government to step in? There is already a taxpayer-subsidized Citizens Property Insurance Corporation run by the State of Florida itself.[71] The government-sponsored insurance company saw a dramatic increase in policyholders: from 700,000 to over a million in a single year. For losses above what their reserves cover, Florida taxpayers will pay. The grim reality is that, sometimes, a town doesn't have to be underwater to become uninhabitable. All it has to do is be uninsurable.[72]

Older homeowners may believe that their insurance policies will cover the damage from hurricanes and rising sea levels. But homeowner policies generally don't cover flood damage, so homeowners living in a flood zone may want coverage from the FEMA, which steps in after disasters.[73] But FEMA was not developed or funded with the expectation of flooding during a period when climate change has changed our expectations. After Hurricane

Katrina and other disasters, homeowners often learned that FEMA would not be there to make them whole after the disaster happened.

One example of the insurance challenge is the case of a homeowner in Louisiana, Tommy Becnel. He had never had a flood claim for his house, but his annual insurance premium is expected to jump ten-fold, up to $7,000 annually in years to come. Becnel, like others, is worried that the new higher insurance price could put their financial security, and their homes as well, at risk. At the same time that private companies are raising their rates, FEMA is raising the rates for the 5 million policyholders it covers. The National Flood Insurance Program covers $1.3 trillion in facilities around the country: the major source for residential flood insurance in the USA. Public or private, access to insurance is being weakened because of losses related to climate change.[74]

Florida has the highest proportion of older Americans. It is also ground zero for climate risk. Many older Americans, including my own parents at retirement, moved to Florida, as real estate prices boomed. But a hurricane can wipe it all out overnight.[75] When it comes to insurance the problem just keeps getting worse, with property insurance rates expected to jump up to 60% in some locations. At the same time, insurance companies are fleeing Florida and other climate-vulnerable states, leaving homeowners without disaster coverage when disasters get worse.[76] "Americans are flooding into Florida expecting 'paradise'—but they're not thinking about this one hidden cost that's only been getting worse."[77]

Hurricanes in Florida are nothing new, as older people well know and remember. Hurricane Ian in September 2022 joined a list of hurricanes including Category 5 Hurricane Andrew, along with Charley, Jeanne, Dennis, Wilma, and Irma, with other major hurricanes going on to devastate states along the East Coast of America. In Florida, as in other devastated states, homeowners chose to live in the path of danger. And people in Florida had higher flood insurance coverage than other states. Insurance costs could send a signal to property owners about where to build in the future. But hearing that signal requires people with ears to hear. The west coast of Florida had many new communities developed during a period when there was a long absence of hurricanes. There are many lessons to be learned from Florida floods, but the most important lesson is that it isn't just about Florida. The Florida flood insurance problem doesn't just stay in Florida but has impacts on the rest of the country.[78]

Hurricane Ian caused more than $50 billion in property damage. Its impact was bigger than any other disaster in U.S. history except for Hurricane Katrina. Along with property damage, Hurricane Ian also devastated insurance companies in Florida, leaving them, like senior citizens, struggling to survive financial losses. Despite Republican free-market ideology, Florida's state legislature and governor made far-reaching changes that gave a subsidy to the insurance industry.[79] In Florida, and other places, what happens is, in climate-risky places, we all pay the price.[80]

When it comes to climate-driven floods, Florida is in a category of its own. Damages since 2017 have equaled nearly 4% of Florida's annual gross domestic product. When all property destruction, infrastructure spending, and power outages are included, Hurricane Ian left more than a 10% hit to the economy. The average annual property insurance payment in the state is $4,231, nearly triple the national rate of $1,544, according to the Insurance Information Institute, an industry association.[81]

Flooding and lack of flood insurance are not limited to Florida. In 2023, California's long-term drought was interrupted by storms and flooding. The result? More than 98% of homeowners lack flood insurance.[82] For fire-prone areas, the same thing has happened: "The marketplace has collapsed," someone observes.[83]

During much of the 20th century, coal mines kept canaries or other creatures to serve as a warning signal for deadly gases. It is reasonable to see homeowners insurance today as the canary in the coal mine for climate change.[84] The question is how to bring the future pain of climate change into the present so that people are motivated to act before it's too late. Insurance could be part of the solution, provided those with enough money can hear the signal of the canary in the coal mine.

What does the canary signal sound like in this case? Homeowners' insurance bills on average cost $1,900, but in areas on the climate-change front line, the bills are much bigger: $4,000 a year in New Orleans and $5,000 a year in Miami. Those who live far from the front line can reasonably say that these places are the canary in the coal mine for climate risk and aging in place. "Doesn't insurance cover that" is a question with a distressing answer: yes, but only at rapidly escalating prices. Claims have increased six times since the 1980s, and insurance companies are waking up and raising premiums. Fire and flood are insurable risks, but the prices go up.

There is a profound unfairness when those who profited least from fossil fuels—the poorest—end up paying for the damage.[85] What we see is that climate risk is borne by lower-income groups and racial minorities but also by older people whose net worth is tied up in their homes and who may be less able to pick up and move somewhere else. Coastal areas and flood zones are joined by those living at the wildland–urban interface or anywhere subject to wildfire risk. Climate change in an aging society will require far-reaching new thinking about insuring for climate risks.[86] In the following chapter, we look at risks that are less insurable at all, like heat waves and drought. Whatever the signal, we need to pay attention to climate risk for aging in place.

Notes

1 Jeff Goodell, *The Water Will Come: Rising Seas, Sinking Cities, and the Remaking of the Civilized World* (Back Bay Books, 2018).
2 *Climate Change Adaptation in Delta Cities* at: www.c40.org/wp-content/uploads/2022/02/C40-Good-Practice-Guide-Climate-Change-Adaptation-in-Delta-Cities.pdf See the PBS series "Sinking Cities" at:

3 "Coastal Populations Set to Age Sharply in the Face of Climate Migration" at: https://phys.org/news/2024–01-coastal-populations-age-sharply-climate.html

4 "Climate Migration Will Leave the Elderly Behind" at: https://grist.org/migration/climate-migration-sea-level-rise-elderly-aging-florida/

5 https://coastal.climatecentral.org/

6 Susan Elizabeth Turek, "Scientists Warn Billions of People Could Soon Be Displaced from 'Climate Niche' Areas That Best Support Life: 'Clear, Profound Ethical Consequences'" *Cool Down* (Mar. 23, 2024) at: www.thecooldown.com/green-tech/mass-displacement-human-population-study-climate-niche-area/

7 "Americans Flocked to Florida for Low Taxes and Sunshine, But an Under-the-Radar Cost of Homeownership Keeps Rising: Insurance" *Business Insider* (July 29, 2023). See also: Tik Root, "Climate Risks Place 39 Million U.S. Homes at Risk of Losing Their Insurance" *Grist* (Sept. 20, 2023) at: https://grist.org/housing/climate-risks-place-39-million-homes-at-risk-of-losing-their-insurance

8 C.H. Jaynes, "Americans Abandoning Neighborhoods Due to Rising Flood Risk, Study Finds" *EcoWatch* (Dec. 19, 2023) at: www.ecowatch.com/americans-climate-migration-flood-risk.html

9 "Climate Shocks Are Making Parts of America Uninsurable. It Just Got Worse" *N.Y. Times* (May 31, 2023) at: www.nytimes.com/2023/05/31/climate/climate-change-insurance-wildfires-california.html Robert Hubbell put it bluntly: See "Climate Denialism Is Not Insurable" *Today's Edition* (July 14, 2023) at: https://roberthubbell.substack.com/p/climate-denialism-is-not-insurable?utm_source=substack&utm_medium=email

10 "Climate Change Is Disrupting Our Sense of Home" by Paige Vega *Vox* (Apr. 22, 2024) at: www.vox.com/climate/24133013/climate-change-displacement-migration-disruption-home

11 Marianne Schnall, "Trailblazing Oceanographer Sylvia Earle on Why We Must Act Now to Safeguard the Ocean, 'Our Life Support System'" *Forbes* (Dec. 7, 2021) at: www.forbes.com/sites/marianneschnall/2021/12/07/trailblazing-oceanographer-sylvia-earle-on-why-we-must-act-now-to-safeguard-the-ocean-our-life-support-system/

12 https://coastal.climatecentral.org/

13 "Injury Deaths Related to Hurricane Sandy, New York City, 2012" at: https://pubmed.ncbi.nlm.nih.gov/27074115/

14 Merrick, Long Island, NY: "Merrick was relatively unscathed in 2012 by Hurricane Sandy compared with other South Shore towns. And though some homes were damaged . . ." most were remediated quickly." Builders bought the few that weren't, and "new construction now stands in their place." *NY Times* (June 24, 2015) at: www.nytimes.com/2015/06/28/realestate/merrick-ny-a-hamlet-with-a-nautical-flavor.html

15 Jordane Valone, "Flood Insurance Rates Could Increase in Bellmore, Merrick" *Long Island Herald* (Apr. 7, 2022) at: www.liherald.com/merrick/stories/flood-insurance-rates-could-increase-in-bellmore-merrick,139891

16 Erica Goode, "A Wrenching Choice for Alaska Towns in the Path of Climate Change" *NY Times* (2016) at: www.nytimes.com/interactive/2016/11/29/science/alaska-global-warming.html

17 Climate Dreams, May, 2019 at: https://climatedreams.com/

18 Erica Goode, "Climate Change Pushes Towns in Alaska to Wrenching Choice" *NY Times* (Nov. 29, 2016), cited by Kelley Bulkeley, "Climate Change Nightmares: A Sign of Things to Come" at: https://kellybulkeley.org/climate-change-nightmares-a-sign-of-things-to-come/

19 Douglas Brinkley, *The Great Deluge: Hurricane Katrina, New Orleans, and the Mississippi Gulf Coast* (Harper Perennial, 2007).

20 Jed Horne, *Breach of Faith: Hurricane Katrina and the Near Death of a Great American City* (Random House, 2008).
   See also: "Nearly Half of the Individuals Who Died During Hurricane Katrina in 2005 Were 75 or Older" at: www.cambridge.org/core/journals/disaster-medicine-and-public-health-preparedness/article/hurricane-katrina-deaths-louisiana-2005/8A4BA6D478C4EB4C3308D7DD48DEB9AB
21 H.R. Moody, *Ethics in an Aging Society* (Johns Hopkins University Press, 1992).
22 Madeleine de Trenqualye, "'It's Inequality That Kills' Naomi Klein on the Future of Climate Justice" *The Guardian* (Feb. 13, 2023) at: www.theguardian.com/books/2023/feb/13/its-inequality-that-kills-naomi-klein-on-the-future-of-climate-justice
23 Mary Heglar, "After the Storm" in *The World as We Knew It: Dispatches from a Changing Climate*, edited by Amy Brady and Tajja Isen (Catapult, 2022), pp. 139–142.
24 Varshini Prakash and Guido Girgenti, *Winning the Green New Deal: Why We Must, How We Can* (Simon & Schuster, 2020).
25 Margaret Morganroth Gullette, "Katrina and the Politics of Later Life" in *There Is No Such Thing as a Natural Disaster: Race, Class, and Hurricane Katrina*, edited by Chester Hartman and Gregory D. Squires (Routledge, 2006). See also Margaret Gullette, *American Eldercide* (University of Chicago Press, in press).
26 Bob Berwyn, "Sea Level Rise Could Drive 1 in 10 People from Their Homes, with Dangerous Implications for International Peace, UN Secretary General Warns" *Inside Climate News* (Feb. 14, 2023) at: https://insideclimatenews.org/news/14022023/sea-level-rise-migration-peace-security-antonio-guterres/
27 UCCCRN, "How Climate Change Could Impact the World's Greatest Cities" (Feb., 2018) at: www.c40.org/wp-content/uploads/2021/08/1789_Future_We_Dont_Want_Report_1.4_hi-res_120618.original.pdf
28 Anthropocene, "The Great American Climate Migration" at: https://anthropocenealliance.org/wp-content/uploads/2021/08/The-Great-American-Climate-Migration.pdf
   See also: "Flooding Drives Millions to Move as Climate Migration Patterns Emerge" *Associated Press* (Dec. 18, 2023) at: https://apnews.com/article/climate-change-flooding-insurance-migration-rain-a43c10d60d5c4f0b9fe30e0a7fe024c4 See "The Flooding Will Come 'No Matter What'" The complex, contradictory and heartbreaking process of American climate migration is underway, by Abraham Lustgarten (*ProPublica*, Apr. 11, 2024) at www.propublica.org/article/climate-migration-louisiana-slidell-flooding See also: Abrahm Lustgarten, *On the Move: The Overheating Earth and the Uprooting of America* (Farrar, Straus, and Giroux, 2024).
29 World Atlas, "These 8 American Cities Could Vanish By 2100" (2023) at: www.worldatlas.com/cities/these-8-american-cities-could-vanish-by-2100.html
30 Christen Hemmingway Janes, "Sea Level Rise Will Affect 4 Out of 5 Miami Residents, Even Those Living Outside of Flood Zones, Study Says" *EcoWatch* (Oct. 17, 2023) at: www.ecowatch.com/miami-sea-level-rise-projections.html
31 Kiley Bense, "Cities Like Philadelphia Are 'Powerful Tools' for Climate Adaptation" *Inside Climate News* (Aug. 28, 2022) at: https://insideclimatenews.org/news/28082022/with-cop27-approaching-cities-like-philadelphia-are-powerful-tools-for-climate-adaptation/
32 Ibid.
33 Cornell University, "Inside Cornell: Aging in the Age of Climate Change" (2013) www.cornell.edu/video/inside-cornell-aging-in-the-age-of-climate-change
34 Victoria Masterson, "Sea Level Rise: Everything You Need to Know" *Centre for Nature and Climate* (Sept. 29, 2022) at: www.weforum.org/agenda/2022/09/rising-sea-levels-global-threat/
35 Dinah Pulver, "Rising Seas Fueled by Climate Change to Swamp $34B in US Real Estate in Just 30 Years, Analysis Finds" *USA Today* (Sept. 9, 2022) at: www.usatoday.com/story/news/2022/09/08/more-than-34-b-us-real-estate-could-flood-2050-analysis-finds/8009627001/

36 Joy Wiltermuth, "Rising Sea Levels Put Miami, New York at Biggest Risk for Severe and Extreme Flooding at Commercial Properties" *MarketWatch* (Sept. 13, 2022) at: www.marketwatch.com/story/rising-sea-levels-put-miami-new-york-at-biggest-risk-for-commercial-property-damage-11663094074?mod=flipboard_investing

37 "Homeowners Face a \$25trn Bill from Climate Change" *The Economist* (Apr. 11, 2024) at: www.economist.com/briefing/2024/04/11/homeowners-face-a-25trn-bill-from-climate-change

38 Robert Young, "To Save America's Coasts, Don't Always Rebuild Them" *New York Times* (Oct. 4, 2022) at: www.nytimes.com/2022/10/04/opinion/hurricane-ian-coast-rebuilding.html?campaign_id=39&emc=edit_ty_20221004&instance_id=73676&nl=opinion-

39 Joaquim Salles, "Left Behind: What Life Is Like for the Last Residents of Staten Island's Oakwood Beach" *Grist* (Sept. 21, 2022) at: https://grist.org/equity/oakwood-beach-staten-island-buyouts-superstorm-sandy/

40 Brady Dennis, "'It's Like a Death:' What It's Like to Leave One Flood-Prone Community" *Washington Post* (Oct. 25, 2022).

41 Christopher O'Donnell, "Ian Was Lethal for the Elderly and the Chronically Ill" *Tampa Bay Times* (Oct. 21, 2022).

42 Amy Green, "Destructive Path. The Numbers Are Horrific" *Inside Climate News* (Apr. 3, 2023) at: https://insideclimatenews.org/news/03042023/hurricane-ian-destruction/

43 "Medical Examiners in Florida Have So Far Linked 112 Deaths to Hurricane Ian. Almost 60 Percent of Those Were 65 or Older. Chronic Medical Conditions Like Respiratory Illnesses Were Contributing Factors" at: www.govtech.com/em/safety/ian-was-lethal-for-the-elderly-and-the-chronically-ill

44 Robynne Boyd, "The People of the Isle de Jean Charles Are Louisiana's First Climate Refugees—But They Won't Be the Last" *Natural Resources Defense Council* (Sept. 23, 2019) at: www.nrdc.org/stories/people-isle-jean-charles-are-louisianas-first-climate-refugees-they-wont-be-last

45 Jack Lee, "Atmospheric Rivers Hitting California Will Become Even More Intense. Here's How They Work" *SF Chronicle* (Dec. 30, 2022) at: www.sfchronicle.com/weather/article/Atmospheric-rivers-hitting-California-will-become-17686559.php

46 Kaitlyn Trudeau, "Yes, Climate Change Influences Atmospheric Rivers" *The Hill* (Mar. 17, 2023) at: https://thehill.com/opinion/energy-environment/3905401-yes-climate-change-influences-atmospheric-rivers/

47 Jake Brittle, "How Rising Temperatures Are Intensifying California's Atmospheric Rivers" *Grist* (Mar. 15, 2023) at: https://grist.org/extreme-weather/climate-change-atmospheric-river-pineapple-express-california-snowpack/

48 Climate Dreams at: https://climatedreams.com/ (Mar., 2019).

49 Clarissa Pinkola Estes, "We Were Made for These Times" at: www.awakin.org/v2/read/view.php?tid=2195

50 From Ronald M. Berndt and Catherine H. Berndt, *The Speaking Land: Myth and Story in Aboriginal Australia* (Inner Traditions 1994). Retold by Allan Chinen, *The Muses of Truth and Transformation: Timeless Tales for Troubling Times* (Routledge, 2024). I am grateful to Dr. Chinen for sharing this tale. See also his books on fairy tales for later life: A. Chinen, *In the Ever After: Fairy Tales for the Second Half of Life* (Chiron, 2018).

51 Number billion-dollar climate-related events in the USA and annual costs of events, based on NOAA's National Centers for Environmental Information.

52 Lois Parshley, "As Climate Risks Mount, the Insurance Safety Net Is Collapsing" *Grist* (Oct. 10, 2023) at: https://grist.org/economics/as-climate-risks-mount-the-insurance-safety-net-is-collapsing/

53 Arwa Mahdaw, "For Some US Residents, It Is Now Impossible to Get Home Insurance—and All Because of the Climate Crisis" *The Guardian* (May 31, 2023) at: www.theguardian.com/commentisfree/2023/may/31/for-some-us-residents-it-is-now-impossible-to-get-home-insurance-and-all-because-of-the-climate-crisis See also: "California Insurance Market Rattled by Withdrawal of Major Companies" at: https://apnews.com/article/california-wildfire-insurance-e31bef0ed7eeddcde096a5b8f2c1768f

54 "Uninsurable America" *Axios* (June 6, 2023) at: www.axios.com/newsletters/axios-am-6d6331d1-d8d5-41f2-acb4-a0d3a81efceb.html See also: "The Costs of Climate Change Are Already Here: In Your Homeowners Insurance Bill" at: www.investopedia.com/the-costs-of-climate-change-are-already-here-in-your-insurance-bill-8414294

55 From: "24 Predictions for 2024" *Grist* (Jan. 3, 2024) at: https://grist.org/culture/24-predictions-for-2024-climate-trends/

56 See www.floodsmart.gov/

57 See Christopher Flavell, "The Cost of Insuring Expensive Waterfront Homes Is About to Skyrocket" *NY Times* (Sept. 24, 2021) at: www.nytimes.com/2021/09/24/climate/federal-flood-insurance-cost.html

58 "Ian Was a Warning to Make Flood Insurance Accessible to All" at: https://thehill.com/opinion/energy-environment/3678420-ian-was-a-warning-to-make-flood-insurance-accessible-to-all/

59 CNBC, "Climate Change Is Making Some Homes Too Costly to Insure" at: https://apple.news/AlW6jpJgkQc-pRU_yEbx31Q

60 Brianna Sacks, "Insurers Slashed Hurricane Ian Payouts Far Below Damage Estimates, Documents and Insiders Reveal" *Washington Post* (Mar. 11, 2023) at: www.washingtonpost.com/climate-environment/2023/03/11/florida-insurance-claims-hurricane-ian

61 "Climate Change Is Hurting Insurers—Report" at: www.yahoo.com/finance/news/climate-change-hurting-insurers-report-041111184.html

62 Eugene Linden, "Insurers Could Have Been Climate Heroes. Instead, They Have Risked a Crisis to Dwarf 2008" *The Guardian* (May, 2022) at: www.theguardian.com/commentisfree/2022/mar/27/insurers-could-have-been-climate-heroes-instead-they-have-risked-a-crisis-to-dwarf-2008

63 Grist, "Homes in Flood Zones Are Overvalued By Billions, Study Finds" *Grist* (Mar. 15, 2023) Failure to account for climate change means low-income homeowners could see their home values plunge. www.ecowatch.com/homes-flood-zone-overvalued-by-billions.html Harress, Christopher, "Insurance Companies Are Fleeing Climate-Vulnerable States, Leaving Thousands Without Disaster Coverage" *Reckon* (Mar. 6, 2023) at: www.reckon.news/news/2023/03/insurance-companies-are-fleeing-climate-vulnerable-states-leaving-thousands-without-disaster-coverage.html

64 Patrick Cooley, "Home Insurance Was Once a 'Must.' Now More Homeowners Are Going Without" *The Washington Post* (May 27, 2024) at: www.washingtonpost.com/business/2024/05/27/home-insurance-dropped-coverage/

65 "Accounting for Flood Risk Would Lower American House Prices by $187bn" *The Economist* (Apr. 10, 2023).

66 Carolyn Kousky, *Understanding Disaster Insurance: New Tools for a More Resilient Future* (Island Press, 2022). "Home insurers are charging more and insuring less" *The Wall Street Journal* (July 30, 2023) at: www.wsj.com/articles/home-insurers-are-charging-more-and-insuring-less-9e948113

67 Leslie Kaufman, "US Housing Market Is Overvalued by Billions Due to Ignored Flood Risk" *Bloomberg* (Feb. 16, 2023) at: www.msn.com/en-us/money/real-estate/us-housing-market-is-overvalued-by-billions-due-to-ignored-flood-risk/ar-AA17zGT9

68 Noor Zainab Hussain and Carolyn Cohn, "Risky Business: Climate Change Turns Up the Heat on Insurers, Policyholders" *Reuters* (Nov. 11, 2021) at: www.reuters.com/business/cop/risky-business-climate-change-turns-up-heat-insurers-policy-holders-2021-11-11/ (Hussain & Cohn, 2021).
69 Christopher Flavelle, "Hurricane Ian's Toll Is Severe. Lack of Insurance Will Make It Worse" *NY Times* (Sept. 29, 2022) at: www.nytimes.com/2022/09/29/climate/hurricane-ian-flood-insurance.html?smid=nytcore-ios-share&referringSource=articleShare
70 Alexis Christoforous, "Hurricane Ian Could Cripple Florida's Home Insurance Industry" *Politico* (Sept. 29, 2022), *ABC News* (Sept. 29, 2022) at: https://abcnews.go.com/Business/hurricane-ian-cripple-floridas-home-insurance-industry/story?id=90638752
71 Florida's Citizens Property Insurance Corp at: www.citizensfla.com/who-we-are
72 Zoe Schlanger, "The Hypocrisy at the Heart of the Insurance Industry" *The Atlantic* (Oct. 18, 2023) at: www.theatlantic.com/science/archive/2023/10/climate-change-home-insurance-companies/675681/
73 But there are problems with FEMA: As FEMA struggles to keep up with climate disasters, extremist groups see an opportunity. See "Boots on the Ground" at: https://grist.org/extreme-weather/boots-on-the-ground-fema-oath-keepers-natural-disaster/
74 David Sherfinski and Diana Baptista, "As Climate Risks Rise, Flood Insurance Costs Stun US Homeowners" *Context* (May 10, 2023) at: www.context.news/climate-risks/as-climate-risks-rise-flood-insurance-costs-stun-us-homeowners
75 Thomas Frank and Daniel Cusik, "Profit Drove a 30-Year Boom. Ian Smashed It in a Day" *Politico* (Sept. 29, 2022) at: www.politico.com/news/2022/09/29/ian-ravaged-one-of-the-fastest-growing-areas-in-the-u-s-00059368
76 Christopher Harress, "Insurance Companies Are Fleeing Climate-Vulnerable States, Leaving Thousands Without Disaster Coverage" *Reckon Media* (Mar. 6, 2023) at: www.reckon.news/news/2023/03/insurance-companies-are-fleeing-climate-vulnerable-states-leaving-thousands-without-disaster-coverage.html
77 "Much of Florida Just Basically Became Worthless" at: https://moneywise.com/insurance/home/americans-moving-to-florida-face-hidden-cost
78 Gareth McGrath, "Why What Happens in Florida Doesn't Stay in Florida: What Ian Means for NC Insurance Rates" *USA Today Network* (Oct. 10, 2022) at: www.starnewsonline.com/story/news/2022/10/07/hurricane-ian-nc-insurance-rates/69534137007
79 Jake Bittle, "Why Republicans Are Coughing Up Billions of Dollars to Save Florida's Insurance Market" *Grist* (Dec. 19, 2022) at: https://grist.org/extreme-weather/florida-insurance-special-session-hurricane-ian-reinsurance-desantis-republicans/?utm_medium=email&utm_source=newsletter&utm_campaign=daily
80 Jenny Schuetz, "Home Mortgage and Insurance Systems Encourage Development in Climate-Risky Places, and We All Pay the Price" *Brookings* (Mar. 9, 2022) at: www.brookings.edu/blog/the-avenue/2022/03/09/home-mortgage-and-insurance-systems-encourage-development-in-climate-risky-places-and-we-all-pay-the-price/
81 By Leslie Kaufman and Tim Quinson, "Five Ways Hurricanes Increase Your Expenses in Florida" *Bloomberg Green* (May 23, 2023) at: www.bloomberg.com/news/articles/2023-05-23/five-ways-hurricanes-make-hurricanes-more-expensive-in-florida
82 Blanca Begert, "California's Storms Are Almost Over. Its Reckoning with Flood Insurance Is About to Begin" *Grist* (Jan. 18, 2023) at: https://grist.org/extreme-weather/californias-storms-are-almost-over-its-reckoning-with-flood-insurance-is-about-to-begin
83 Christopher Flavelle, "Wildfires Hasten Another Climate Crisis: Homeowners Who Can't Get Insurance" *N.Y. Times* (Sept. 2, 2020) at: www.nytimes.com/2020/09/02/climate/wildfires-insurance.html

84 Benjamin Keys, "Your Homeowners' Insurance Bill Is the Canary in the Climate Coal Mine" *N.Y. Times* (May 7, 2023) at: www.nytimes.com/2023/05/07/opinion/climate-change-homeowners-insurance-housing-market.html
85 "Climate Risk: Insuring Against the Inevitable" at: www.dw.com/en/climate-risk-insuring-against-the-inevitable/a-46615364
86 Sueellen Campbell, "Readings on Climate-Change Impacts and Insurance" *Yale Climate Connections* (May 13, 2022) at: https://yaleclimateconnections.org/2022/05/readings-on-climate-change-impacts-and-insurance/ See also: "Rethinking Insurance for Floods, Wildfires and Other Catastrophes" at: https://knowablemagazine.org/article/society/2022/rethinking-insurance-floods-wildfires-other-catastrophes?utm_source=pocket-newtab

Part II

Health and Age in a Warming World

4 Heat Wave
The Heat Is On

Heat waves are an increasing part of our climate future.[1] Climate experts agreed that 2023 was the earth's hottest year: the warmest since records were kept and perhaps the warmest in 125,000 years. "The rate of warming over the past century has no precedent as far back as we are able to look, not only hundreds or thousands, but many millions of years," according to University of Pennsylvania meteorologist Michael Mann."[2]

Higher heat means death. Between 2014 and 2018, there were more than 10,000 heat-related deaths in the USA, according to the CDC.[3] With global warming the problem will get worse and population aging means that older people will be among the biggest victims. This threat cuts across generations: that is, we who are old and younger people who want to grow old—we all must prepare for life on a scorching planet.[4]

Climate change in an aging society is already presenting clear age-specific threats to health:

> Older adults face higher risks from climate change compared to the rest of the population. For example, older adults are vulnerable to being trapped in poor environments through lack of mobility, disability, and frailty. They are at increased risk of heat-related illnesses, compounded by living alone, co-morbidities, medication, and are at higher risk of dehydration than young people, due to the physiological changes that occur as part of the ageing process. Consequently any reduction in access to fresh water, during a drought for example, has severe implications for older adults. Even mild dehydration adversely affects mental performance, memory, attention, concentration and reaction time, weakness, dizziness and increased risk of falls. Kidney function, cardiac function and convulsions may follow from acute dehydration. Furthermore, drinking contaminated water leads to many diarrheal diseases including dysentery, Hepatitis E, cholera, and typhoid which are frequently fatal in older people.[5]

Our current climate crisis has been concisely identified by the Climate Reality Project: triple-digit temperatures putting stress on the power grid;

DOI: 10.4324/9781003345992-6

wildfires shutting down highways; and severe drought causing water cuts. Record-breaking temperatures are now becoming the new normal. By the mid-21st century, we can expect over 100 million Americans to be experiencing an "extreme heat belt," which is 13 times the number who encountered such extreme temperatures in 2022. What happens when 20% of America's population is over age 65? In coming decades, it is clearly predictable that, along with rising sea levels, we will see rising temperatures and a threat to the health of people over the life course.[6]

Global warming has already begun and has affected land and sea, including the USA. A third of Americans are already seeing above-average warming, but the impact varies depending on locality. The overall picture is now clear. Experts believe that the 1.5°C increase could be surpassed by 2030 unless there are substantial cuts in emissions of carbon dioxide. That 1.5-degree threshold was identified in the 2015 Paris Climate Agreement as a point where we could anticipate severe storms, flooding, and heat waves, which are the focus of this chapter.

We are now living in a warming world: "The warming we are seeing is pushing at the bounds of lived human experience, of what we thought was possible," said Katharine Heyhoe, Chief Scientist of the Nature Conservancy environmental group. She went on: "We are paying the costs for that and we need to prepare for the changes already set in motion, as well as to prevent further warming."[7]

Heat kills more people than any other extreme weather event on the earth, and that includes all the hurricanes, tornadoes, wildfires, and other threats that make the headlines. No wonder we want air conditioning, discussed in more detail in the following text.

The bad news is that heat waves are in our future, and here a picture is worth a thousand words.[8] Why is this important for an aging society? Because as we get older, the body's ability to respond to heat weakens and that means older people are at greater risk for heat-related illnesses.[9] Over 90% of excess deaths in heat waves occur in the elderly, and these deaths are mostly cardiovascular in origin.[10] "It's not just heat stroke. Extreme temperatures pose special risk to people with chronic illness (and that's a lot of us)," said Isabella Cueto.[11]

The editors of over a dozen health journals from across the globe simultaneously published a joint editorial calling for urgent climate action to avert catastrophic warming. Without it, the editorial said, rising temperatures will lead to more deaths from heart and lung illness, allergies, kidney problems, and pregnancy complications. America is vulnerable just like other countries: "The greatest threat to global public health is the continued failure of world leaders to keep the global temperature rise below 1.5°C and to restore nature," said *The New England Journal of Medicine*, in a series focused on highlighting health hazards linked to planet-warming pollution.

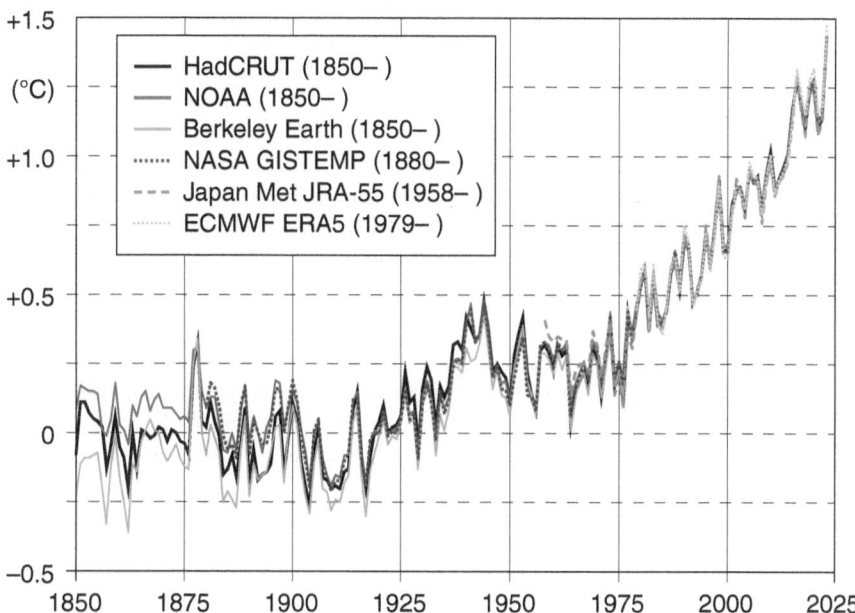

Figure 4.1 Global average temperature change from several scientific organizations. Public domain.

Impact of Heat With Age

Aging as a biological process can be defined as diminished reserve capacity, so it is no surprise that older people will be more likely than younger people to die from heat stroke. Diminished reserve capacity also means that extreme heat or smoke can aggravate heart disease, diabetes, heart disease, and lung conditions. Aging may be different in each person, but the overall pattern is one of increasing vulnerability and higher susceptibility to disease.

In a warming world, we can expect to see heat harming people of all ages. We face not just heat exhaustion or heat stroke but conditions that send people to emergency rooms for a variety of reasons: increased likelihood of work injuries, heart attacks, or aggravated chronic conditions such as chronic obstructive pulmonary disease, migraine, kidney failure, and many more. The body copes with heat by sweating, but sweating causes other side effects, such as dehydration, a factor in other heat-related conditions, including blood glucose levels, blood pressure, or kidney function. For older people, the threat is clear: in recent decades, the number of heat-related deaths among people over 65 has increased by 50%.[12]

Older adults are seriously at risk in very specific ways:[13]

Older adults are more vulnerable to extreme weather events due to aging immune systems and propensity for dehydration because of their decreased sense of thirst and lower body fluid content. Older adults with chronic health conditions such as cardiovascular disease, hypertension, obesity, type 2 diabetes, and chronic kidney disease are especially at risk for disease exacerbations during times of temperature extremes, high stress, and limited resources. Elders with dementias (i.e., Alzheimer's disease, vascular dementias) are at high risk during high temperatures for exacerbation of cognitive decline and hospitalizations.

How much can we expect to adapt to climbing temperatures? There are serious limits to how much heat the human body can take.[14] There are common symptoms of heat-related illnesses including throbbing headaches; dizziness; hot or dry skin; body temperature of 104°F or higher; decreased sweat and tear production; rapid heartbeat; difficulty walking; confusion; seizures.[15] Note that some of these symptoms—like confusion or difficulty walking—may not be probable for compromised individuals in heat waves. Advice like "stay in an air-conditioned building" does not make sense if the air conditioning isn't there at all. "Don't rely on a fan as your main cooling source" doesn't work if a fan is your cooling source. If climate mitigation fails, we will need to think more deeply about adaptation for global warming for an aging society. Adaptation by individuals may not be an approach that works.

Population aging is happening at the same time that climate change is accelerating. Both trends will be costly. Heat waves have the biggest impact on people with chronic health conditions, which describes the older population. Nearly three-quarters of total health care spending goes for chronic diseases[16] but such spending mostly means responding to symptoms, not mitigation or adaptation for climate change. Without mitigation—that is, reduction—of global warming, millions of lives are at stake.[17]

Heat Waves in a Warming World

Heat waves are coming. Increased fossil fuel emissions have raised global temperatures and heat waves will be more frequent and more intense.[18] Today global temperatures are already 2 degrees Fahrenheit above what they were a century and a half ago. Our body temperature remains the same, but ambient temperature is an enormous threat. The World Meteorological Organization has concluded that climate change will push global temperatures to record high levels in coming years.[19] The heat waves are coming at a time when 10,000 aging Boomers turn age 65 each day.

General warming predictions are well-supported, but recent heat waves are a stress test: Do these heat waves mean climate change is happening faster than expected?[20]

When we think about heat waves as a threat in a warming world we usually think about distant places: Africa, the Middle East, and so on. In the USA, we think of heat waves in places like Arizona or California, home of Death Valley. But one of the biggest heat waves—devastating to older people—took place in Chicago.

Heat waves remain a persistent threat in many parts of the world. For instance, in April 2023, a heat wave in India caused the death of at least 11 from stroke. Another 50 were hospitalized. It was another warning to a country vulnerable to the threats from global warming. In this case, it couldn't even be described as a heat wave because it didn't qualify for the technical definition. Moreover, humidity, as well as temperature become part of the threat. Still, in 2022, India registered its hottest month of March since it began keeping records 120 years ago. The major city of Delhi recorded 104°F.[21] Scientists have concluded that climate change has made such heat waves 30 times more likely.[22] The Chicago Heat Wave of 1995 suggests that the USA is vulnerable to such heat waves in the future, too, and age is likely to be a factor in mortality risk.[23]

4.1   Robin Wall Kimmerer: Eco-Elder

Robin Wall Kimmerer, born in 1953, has brought the voice of Indigenous peoples to the challenge of climate change. She is a botanist and specialist in forest biology, but, most significantly, she is an enrolled member of the Citizen Potawatomi Nation and he brings a Native American heritage to environmental issues that are Inescapable in a period of climate change.

Her book, *Braiding Sweetgrass* has made her dual perspective available to Americans of all backgrounds. Kimmerer has worked on the natural history of mosses, but she has also worked on ecology education and incorporated traditional ecological knowledge into that work. Her work earned her a MacArthur Award for creativity and innovation.

Robin Kimmerer is a cross-cultural emblem of what an Eco-Elder can be in our time.

The Big Heat Wave Has Come

Uttar Pradesh, a state with 200 million inhabitants in Northern India, has known heat waves. But nothing like this heat wave had happened before: "It just kept getting hotter," said one of the few survivors. For weeks both the temperature and humidity soared off the charts, until eventually the electricity grid collapsed. We saw Dante's Inferno in the 21st century. Anyone in Uttar Pradesh would have heard cries from the city's rooftops. People were desperately trying find cool night air, only to wake up and find old, sick or younger family members dead from the heat.

Some people went to nearby lakes, but the water was too warm to give any relief. Others tried to find places with air-conditioning, but the electricity had gone out and local generators never worked long enough. Armed men tried to steal any air conditioning units and generators where they could find them. People were dying both day and night as the "wet bulb" temperature, humidity, rose above what human beings could survive. By the end of the week 20 million inhabitants of Uttar Pradesh were dead.

What you've just read here is fiction. It's from the opening pages of Kim Stanley Robinson's book *The Ministry for the Future*. This imaginary scene might have taken place in 2030, but not far off. Is it merely science fiction? Kim Stanley Robinson said this about the scene: "I knew the opening scene would be hard to read, and it was hard to write."

Bill Gates said this about this scene: "It's as harrowing a scene as any I've read in a science fiction book—because the events depicted in it could very well take place in the real world. I don't think we're going to experience heat waves precisely as long or as severe as the one in the book over the next few years. But if we don't decisively reduce carbon emissions and eventually eliminate them, in the decades ahead we could very well see successive days when there's a deadly combination of extremely high temperatures and high humidity."

The *Ministry for the Future* is fiction—"cli fi" or climate fiction. But our own nightmares can display the same threat, as happens in science fiction films[24] and as happened to me in my own dream I had.

Driving to Disaster

The night of Thanksgiving I had been talking to my wife about Kim Stanley Robinson's book *The Ministry for the Future*. In my dream my wife and I were coming from a climate advocacy event and driving through the city in the dark. I couldn't see ahead and eventually I tried to turn on the headlights. But the lights didn't work at all, so I pulled over to the side to get out to see how to fix it. Then I woke up.

The night before this dream my blood pressure hit its highest level of the week. That day I'd been in personal contact with John Berger, author of *Solving the Climate Crisis: Frontline Reports from the Race to Save the Earth* (Seven Stories Press, 2023) and Max Wilbert, coauthor, *Bright Green Lies: How the Environmental Movement Lost Its Way and What We Can do About it* (Monkfish, 2021).

Waking or sleeping, each of us will try to do what we can. So we show up, as I did in my dream showing up for the climate event. Each of us is driving through the dark city of today, trying to "turn on the lights" to see what the future may bring. Are we waiting for 2030? 2045?. But our "headlights," our future forecasts, don't work well. In the dream I tried as hard as possible to turn on the lights to see what is ahead, just as the Intergovernmental Panel on Climate Change (IPCC) does year after year. When we're each driving our separate individual cars, it seems like we're in charge, as if we have agency, just as in the dream I tried to grab the steering wheel. But our agency as individuals is limited. We need to "get out of the car" to find out what is wrong with our situation and "what we can do about it." Otherwise, we're driving to disaster. We need agency with others, not alone.

In July 1995 a heat wave took place in Chicago over a five-day period. When it was over 739 heat-related deaths were recorded, mostly older people. But the numbers could be higher. During that same period the heat wave was felt in other places throughout the Midwest. In Chicago it was severe: the highest temperatures were recorded between July 12 and July 16, reaching 106°F. Nighttime temperatures were also historically high. What made the Chicago heat wave especially damaging was that the wet-bulb temperatures, including humidity, way beyond normal, at national record levels.

Eric Klinenberg, author of *Heat Wave: A Social Autopsy of Disaster in Chicago,* pointed out that heat-related deaths in Chicago could be directly mapped onto the distribution of poverty. The mostly elderly victims were people who could not afford air conditioning. They were often reluctant to open windows or sleep outside because of fear of crime. Further inquiry confirmed that Black and Hispanic people were more likely than whites to die in the heat wave. Authorities were slow to respond to the disaster, so no official death toll was ever established. The figure of 739 heat-related deaths reflects the likely figure from the best epidemiological calculations. The 1995 Chicago wave was a disaster, but it was not the worst heat wave in U.S. history.

The Chicago heat wave also set new records for electricity use. Just as in the Uttar Pradesh fictitious scene, heat wave led to failure of power grids: at one point, 49,000 households in Chicago lost electricity. As in Uttar Pradesh inhabitants congregated on city beaches or used fire hydrants: over 3,000 fire hydrants were opened so that some people lost water pressure. The most vulnerable of all groups were older people, especially those living alone. The Chicago heat wave was much more than just normal warm weather. On its first day, the temperature reached 106°F and subsequently ranged between the 90s and low 100s. But the "wet bulb" impact—heat plus humidity—made it much worse.[25]

The human body cannot cope with continuous heat beyond a certain level, and after two days, heat-related illnesses began to appear in Chicago hospitals. Soon 23 hospitals were on bypass status, closing entry to their emergency rooms. Even the municipal autopsy facility became overcrowded, as refrigerated trucks brought in corpses.

What you just read about Chicago is NOT fiction. Unlike the scene about Uttar Pradesh, the Chicago heat wave happened closer to home. It didn't happen in Death Valley, California but in one of the largest cities in America and a northern state. But other historically unprecedented heat waves did take place in 2022 in the USA and heat waves of this magnitude have taken place in recent years all across the globe. We are living in a time when more heat waves are predictable.

Eric Klinenberg said about the 1995 heat wave: "Chicago's houses and apartment buildings baked like ovens." Then was it all just a "natural disaster?" Klinenberg replied: "Hundreds of Chicago residents died alone, behind locked doors and sealed windows, out of contact with friends, family, and neighbors, unassisted by public agencies or community groups. There's nothing natural about that." Chicago was not alone that year since the heat wave also had its impact on St. Louis and Milwaukee, Wisconsin.

On the positive side, Klinenberg notes that Chicago made management changes and dramatically reduced the number of deaths in the 1999 heat wave. In short: Adaptation but not mitigation. As Klinenberg put it: "We know that more heat waves are coming. Every major report on global warming—including the recent White House study—warns that an increase in severe heat waves is likely." We are living in a time when heat waves are the new normal and where elders are at risk.

The Chicago heat wave was mainly the result of high humidity. That point was made in the fictional scenario of Uttar Pradesh, depicted in *The Ministry of the Future*. The key point is that in muggy, humid air, the human body will struggle to cool off, because evaporation by sweat doesn't work as well. Hence, the importance of "wet-bulb" measurements for assessing heat waves. Temperature alone is not the final predictor of pathology, which is why a heat index—like that provided by the National Weather Service—may not adequately measure the threat.[26]

We know about the physiology of heat and metabolism, so it is not surprising that thousands of elder residents experienced heat-related symptoms. Just as in the COVID pandemic decades later, the health care system including hospitals and ambulances, couldn't keep up with the pressure and growing numbers couldn't get access to emergency care. When the Centers for Disease Control analyzed risk factors of heat wave victims their list included the ones we might expect: living alone, lack of access to transportation, social isolation and, above all, not having air conditioning.

Klinenberg's analysis of the Chicago heat wave led him to identify factors that are important as we think about possibilities for adaptation for heat waves in an aging society. The Chicago urban environment was a condition where people did not trust neighbors and were cut off from services of all kinds. What came to be called "precarity" is a good description of their condition, which means that, whatever individual autopsies might say, the heat wave was a social disaster reflecting inequality, much as we would later see during the COVID pandemic.[27] Klinenberg favors investment in "social infrastructure" as the path to overcome this vulnerability.[28] His solution is exactly what is needed to prepare for adaptation for heat waves in an aging society.

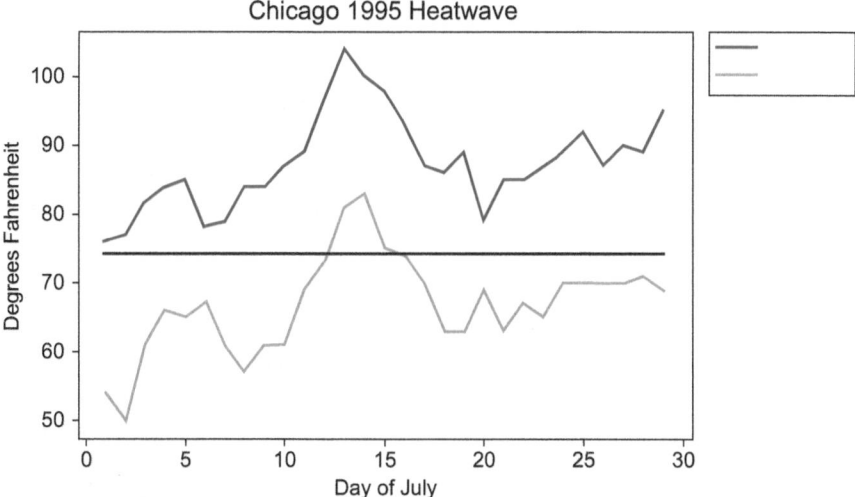

Figure 4.2 Observed temperatures at Chicago O'Hare Airport. Red Lines are daily maximum. Blue Lines are daily minimum; Black Line is July 1960–2016 mean temperature. Public domain.

Extreme weather events include extreme heat waves, and these are becoming more common in a warming world. Record high temperatures have been recorded across the globe: 112°F in North Texas, 97°F in Minneapolis, 103°F in Nebraska, and 104°F in London.

From 1998 to 2017, the World Health Organization estimates 166,000 people died from heat waves globally, and that is likely an undercount. The problem is getting worse, too. Exposure to extreme heat has tripled in the last few decades, and now afflicts nearly a quarter of people on earth, an analysis by the Associated Press found. Heat kills more people each year, an estimated 1,300 Americans, than any other weather-related event. By 2050, that number could be closer to 60,000 deaths each year.[29]

Naturally, there is hope that greater public recognition of the threat of heat waves will lead to action: for example, Miami has, for the first time, appointed a "chief heat officer," in response to climate issues.[30]

Can We Get Used to Warmer Weather?

Over recent decades heat waves in America have gotten worse: hotter, longer, and more frequent. In the 1960s heat waves could be counted around two a year, but, by the 2010s, heat waves were up to ten per year, according to the National Oceanic and Atmospheric Administration. How long do heat waves last? The heat wave season each year has more than tripled: in the same period, the length has grown from 22 to 68 days, a huge increase. What kills are extremes of heat, and extremes are now found in geographic regions completely inexperienced with such excessive heat. One example is the Pacific Northwest, which saw hundreds of deaths from a heat wave in 2021.[31]

We are used to thinking of temperature as purely numerical: "The high will be 90 degrees today." But the temperature of a warming world cannot be reduced to numbers alone. Culture and biology—including age—will be important in how that warming world is experienced.

There are many different numerical measurements that try to track what the warming world will look like: heat index is one of these; the web bulb global temperature is another. We know that indexes may need to be recalibrated if people move from one temperature zone to another. But what happens if in a limited period—a decade, for example—global warming is experienced by people who have grown old under different conditions? What happens when "the climates people have acclimatized to over their whole lives have become much hotter and wetter"?[32]

Empirical studies support the idea that where someone grew up will affect the temperature received as safe and appropriate.[33] The problem is that this kind of long-range adaptation may mean that older people, for example, feel comfortable at higher temperatures, even if the health risk is greater than they perceive.

In June and July 2021, a heat wave in western North America caused at least 868 deaths, triggering wildfires and flood warnings from rapid snow and ice melt. They affected parts of California, Idaho, Nevada, Oregon, and Washington State, as well as British Columbia.[34] Older people are living in a world different from the world they grew up in.

As the earth's climate warms, however, hotter-than-usual days and nights are becoming more common and heat waves are expected to become more frequent and intense, according to the U.S. Environmental Protection Agency. "Every heat wave in the world is now made stronger and more likely to happen because of human-caused climate change," said Friederike Otto, a climate scientist at Imperial College of London's Grantham Institute.

Research suggests that up to 800,000 residents of Phoenix could need emergency medical care for heat stroke and other heat-related conditions in the event of an extended power failure.[35] But other parts of America are also at risk. A study on the record-breaking heat wave in the Pacific Northwest, for instance, concluded that it was "virtually impossible" that it would have occurred without human-caused climate change.[36]

Air Conditioning

The problem of air conditioning is actually simple: "The warmer the world gets, the more people turn to their air conditioners. And the more they turn to their air conditioners, the warmer the world gets."[37] The paradox is that air conditioning and refrigerators do help keep people healthy and safe—but they also contribute to global warming.

The facts are clear enough. The World Economic Forum estimates that greenhouse gas emissions from air conditioning might account for half a degree Celsius in rising global temperatures during this century. The International Energy Agency estimates that air conditioners and electric fans are the source of 10% of global electricity consumption. Despite these facts, air conditioning is rarely discussed in public discussion about the cause of global warming.

When John Kerry was Secretary of State, at an international conference, Kerry said air conditioning is likely the single most important step we could take to limit the warming of the planet and limit warming for generations to come.[38] Yet, despite its importance, air conditioning is often overlooked. It needs to be a simple important part of becoming a climate-conscious consumer. In an earlier chapter, we discussed the impact of heat waves in the coming years, and access to air conditioning is an urgent issue of public health.[39] In an aging society, it becomes more acute because of vulnerability to heat with advancing age.

The use of air conditioning in an aging society raises questions that deserve serious attention. For example, does it make sense to run an air conditioner all summer long without break, or should people turn it off during the day if they're not using it when there's no benefit from it? This question is not simple when we study the effects of air conditioning: "These effects mean there's no one straightforward answer to whether you should blast the A/C all day or wait until you get back home in the evening".[40] Consider that people who are retired, like the author of this book, typically spend much of the day at home: commuting to work is not an issue. Retirement is only one reason for being at home. More and more people of mid-life are now working remotely, that is, working from home.

Just how big is air conditioning? A lot bigger than we think. Nearly 4% of all greenhouse gas emissions come only from air conditioning. That figure is likely to grow as developing countries in heat zones experience heat waves and are able to afford air conditioning. The International Energy Agency estimates that air conditioning could triple by mid-century.[41]

Today roughly 90% of U.S. homes have some form of air conditioning, according to U.S. Census data. For decades in Europe, leaders and scholars scoffed at U.S. reliance on air conditioning as another example of American excess. In 1992, Cambridge economist Gwyn Prins warned that "physical addiction to air-conditioned air is the most pervasive and least noticed epidemic in modern America." American Geophysical Union is concerned that U.S. household air conditioning use could exceed electric capacity in coming decades because of climate change.[42]

America is different from Europe when it comes to air conditioning. According to one industry estimate, just 3% of homes in Germany and less than 5% of homes in France have air conditioning. In Britain, government

estimates suggest that less than 5% of homes in England have air conditioning units installed.[43]

"I really can't blame Europeans for buying air conditioners right now,"
said climate control researcher Stan Cox.

> Those who lived through the 2003 heat wave that killed tens of thou-
> sands were told that it was a fluke, something that happens once every
> few centuries. Now they're being hit again, just 19 years later, with
> more heat waves likely to come.

Heating and cooling are the largest energy expenses in most homes—comprising 35% to 50% of the annual electricity bill. They also account for a
significant portion of an individual's impact on the climate.

There may be alternatives to air conditioning, from new technology (e.g.,
heat pumps), which are more energy efficient than older technology. But other
alternatives include smaller rooms or stronger insulation. But air conditioning
seems likely to be a part of Europe's future in one way or another. In the USA
in 2010 we used as much electricity just for air conditioning as we used for all
purposes in 1955. From 1993 to 2005 electricity for air conditioning doubled.
More electricity goes to air conditioning than to the construction industry.[44]

Americans use around 17% of annual electricity for air conditioning, but
on the hottest days, and in certain parts of the country, air conditioning can
account for up to 70% of electrical use. Air conditioning is not just a remedy
for symptoms of global warming. It is also a contributor to the problem.
Most air conditioners are still using refrigerants that produce greenhouse gas
emissions vastly more powerful than carbon dioxide.[45]

The need for air conditioning is a global problem:

> Currently, 2.2 billion people—that's 30% of the world's population—
> now experience life-threatening heat during at least 20 days of the
> year, and scientists predict that heat could threaten as many as 66% of
> human lives by the end of the century.[46]

By some estimates, international demand for cooling will triple in the next
three decades. Kim Stanley Robinson begins his important novel *The Ministry for the Future* with an account of a disastrous heat wave in India at a time
when electricity and air conditioning have failed. The story is fiction, but the
threat is already real.

The challenge is clear: both mitigation and adaptation are called for. Eric
Wilson identifies skills of adaptation that individuals can put into practice on
their own: for example,

> cold showers, well-placed fans that encourage vigorous cross-breezes,
> opening windows at strategic places in an apartment or house, drawing
> curtains or shades when direct sunlight hits windows, avoiding alcohol

and hydrating, unplugging necessary appliances that may generate heat, learning to distinguish between heat discomfort and danger, connecting with isolated neighbors, sleeping with wet socks pointed at a fan, and many more.[47]

Time for full disclosure. When I was growing up, in the 1950s, we never had air conditioning. Most people had to go to a local movie theater, The Gables in Merrick, to get cool in the summer. Even today, I have no air conditioning, thanks to comfortable Pacific breezes in San Mateo, California. But people in the Pacific Northwest in the past also had little air conditioning. Then came heat waves there, too, and maybe soon for all of us.

Air conditioning poses a dilemma in an era of climate change in an aging society: air conditioning, like refrigerators, keeps people safe and comfortable. But both of these also warm the planet.[48] And there can be alternatives to air conditioning.[49]

Air conditioning is a great invention, both for economic productivity and for preventing heat-related death. But by the 1980s we learned that air conditioning produces dangerous hydrofluorocarbon gases that damage the ozone layer. Air conditioning also consumes vast amounts of energy from fossil fuels.[50] How can we respond to the need for air conditioning and also find ways to people cool without warming the earth? There are also issues of equity and economics related to air conditioning, as we saw clearly in the impact of the Chicago heat wave in 1995.[51]

Technology researchers and entrepreneurs are working to create more efficient air conditioning systems.[52] One example might be desiccant dehumidifiers for humidity control and more efficient latent load removal to achieve reduction for peak electric demands. Surveys confirm that Americans, far more than Europeans, think that technological breakthroughs will be the solution to climate change.[53] We can be hopeful for technology as we aim for mitigation in response to climate change. But there are alternative, less expensive options to cool down besides conventional air conditioning.[54] Traditional cultures have lessons in adaptation that may need to be relearned.[55]

Along with mitigation, others have come up with adaptation responses. Recognizing the threat of losing electricity for air conditioning, there are tactics for coping with heat even without air conditioning. These steps include drinking liquids, using water in creative ways, scheduling daytime activities, and wearing lighter clothing.[56] Adaptation may produce remarkable effects in years to come, and it often depends on people acting on a small scale in their own individual settings.[57]

What is an Eco-Elder? Above all, Eco-Elders are those who are old and are concerned about future generations. James Hansen expressed it in his book *Storms of My Grandchildren: The Truth About the Coming Climate Catastrophe.*

Television journalist Hugh Downs expressed something similar in his *Letter to a Great Grandson: A Message of Love, Advice, and Hopes for the Future* (Scribner, 2004).

4.2   Hugh Downs: Eco-Elder

Figure 4.3 Hugh Downs. Public domain.

"The Sea is the ultimate answer," said oceanographer Sylvia Earle when she went diving with Hugh Downs in the Key Largo National Marine Sanctuary. Hugh Downs at the time (1992) was cohost of the 20/20 TV show. But television legend Downs was a man of many interests and a deep commitment to planet Earth. In the 1990s I began meeting with him every month or so. He hosted a PBS series about aging, "Over Easy" and he wanted to have a personal tutor to give him a background in gerontology. Meeting him regularly became part of my job at the Brookdale Center on Aging of Hunter College. As I got to know him, I was amazed to learn that he had read the entire Great Books and that he had a passionate interest in earth science and environmental protection.

At the time of the first Earth Day in 1970 the *Arizona Republic* ran an article about the day on its front page and on the Today Show, Hugh Downs dedicated its program to Earth Day for the whole week. At that time he said this:

> Do we have the will to turn our way of life upside down? To make personal, corporate, and national sacrifices in order to keep this earth

alive? Or do we go on breeding, demanding more and more power, more of everything until we suffocate or die of plague or famine?

Hugh Downs challenged viewers on that first Earth Day in 1970: "Our oceans are dying, our air is poisoned. This is not science fiction and it is not the future; it is happening now and we have to make a decision now."[58]

After his retirement, I visited him at his home near Phoenix, Arizona, and discovered that lifelong learning and commitment to the public good continued in full force. It was something he would never retire from. He remained an Eco-Elder until he died at the age of 99 in 2020.

Notes

1 Jeff Berardelli, "Heat Waves and Climate Change: Is There a Connection?" *Yale Climate Connections* (June 25, 2019) at: https://yaleclimateconnections.org/2019/06/heat-waves-and-climate-change-is-there-a-connection/

2 www.usatoday.com/story/news/nation/2024/01/01/2023-was-earths-hottest-year-experts-say/71882923007/ Caution: we should not equate heat waves themselves with global warming:

> Since the late 1800s, temperatures have already risen 1.1 degrees Celsius (2 degrees Fahrenheit). This also sends more moisture from evaporation into the atmosphere, increasing the intensity and frequency of heavy precipitation. It can even cause localized cooling or snowstorms . . .
> Localized cooling doesn't change the fact that the Arctic is warming nearly four times as fast as anywhere else on the planet, and we're headed for a dangerously hot world.

From Michael J. Coren, "The Climate Coach" *Washington Post* (Jan. 4, 2024) at: https://s2.washingtonpost.com/camp-rw/

3 "Heat-Related Deaths—United States, 2004–2018" at: www.cdc.gov/mmwr/volumes/69/wr/mm6924a1.htm

4 Jeff Goodell, *The Heat Will Kill You First: Life and Death on a Scorched Planet* (Little, Brown, 2023).

5 Sarah Harper, "The Convergence of Population Ageing with Climate Change" *Journal of Population Ageing* 2019;12: 401–403 at: https://link.springer.com/article/10.1007/s12062-019-09255-5

6 See: "Deadly Heat Waves Threaten Older People as Summer Nears" at: https://apnews.com/article/climate-aging-heat-deaths-f95f3567fe9fa97c03c59bdfe839e202 See also: "Takeaways about Heat Deaths and Vulnerable Older People" at: https://apnews.com/article/extreme-heat-deaths-aging-climate-47a8089cdecc84a0f19697438975353f

7 Oliver Milman, "A Third of Americans Are Already Facing Above-Average Warming" *The Guardian* (Feb. 5, 2022) at: www.theguardian.com/environment/2022/feb/05/americans-above-average-temperature-increase-climate-crisis?CMP=Share_iOSApp_Other

8 www.statista.com/chart/28722/global-surface-temperature-anomalies/

9 K.B. Pandolf, "Aging and Human Heat Tolerance" *Experimental Aging Research* 1997;23(1):69–105 at: https://pubmed.ncbi.nlm.nih.gov/9049613/ See also CDC: www.cdc.gov/aging/emergency-preparedness/older-adults-extreme-heat/index.html

10 W. Larry Kenney, W. Larry, Daniel Craighead, and Lacy Alexander, "Heat Waves, Aging and Human Cardiovascular Health" *National Library of Medicine* (2014) at: www.ncbi.nlm.nih.gov/pmc/articles/PMC4155032/

11 Isabella Cueto, *Stat* (July 19, 2022) at: www.statnews.com/2022/07/19/heat-waves-risk-to-people-with-chronic-illness/

12 British Medical Journal, "Call for Emergency Action to Limit Global Temperature Increases, Restore Biodiversity, and Protect Health" (Sept. 6, 2021) at: https://doi.org/10.1136/bmj.n1734

13 From: "Climate Change: Effects on the Older Adult" at: www.sciencedirect.com/science/article/abs/pii/S1555415522000071

14 Casey Crownhar, "Here's How Much Heat Your Body Can Take" *Technology Review* (Aug. 3, 2023) at: www.technologyreview.com/2023/08/03/1077155/extreme-heat-health/

15 Jeffrey M.D. Luther, "Senior Living: The Dangers of Heat-Related Illnesses to Older Adults" *Los Angeles Daily News* (June 27, 2022).

16 *The Cost of Chronic Disease in the U.S.*, Milken Institute (Aug., 2018).

17 Sarah Kaplan, "Inaction on Climate Change Imperils Millions of Lives" *Washington Post* (Oct. 21, 2021) at: www.washingtonpost.com/climate-environment/2021/10/20/lancet-climate-inaction-threatens-millions/ See also: "Without Human-Caused Climate Change Temperatures of 40°C in the UK Would Have Been Extremely Unlikely" at: www.worldweatherattribution.org/analysis/heatwave/

18 "Fossil-Fuel Pollution and Climate Change" *New England Journal of Medicine* at: www.nejm.org/doi/full/10.1056/NEJMe2206300?utm_source=newsletter&utm_medium=email&utm_campaign=wp_climate202&wpisrc=nl_climate202

19 Brad Plumer, "Heat Will Likely Soar to Record Levels in Next 5 Years, New Analysis Says" *NY Times* (May 23, 2023) at: www.nytimes.com/2023/05/17/climate/record-heat-forecast.html

20 James Temple, "Do These Heat Waves Mean Climate Change Is Happening Faster Than Expected?" *MIT Technology Review* (July 21, 2022) at: www.technologyreview.com/2022/07/21/1056291/do-these-heatwaves-mean-climate-change-is-worse-than-we-thought/

21 World Weather Attribution Header, "Climate Change Made Devastating Early Heat in India and Pakistan 30 Times More Likely" (May 23, 2022) at: www.worldweatherattribution.org/climate-change-made-devastating-early-heat-in-india-and-pakistan-30-times-more-likely/

22 Somini Sengupta, "Coping with Extreme Heat" *NY Times* (Apr. 18, 2023) at: www.nytimes.com/2023/04/18/climate/extreme-heat.html

23 Joan Meiners, "Phoenix Is Not Prepared for a Simultaneous Heat Wave and Blackout, New Research Shows" *Arizona Republic* (May 25, 2023) at: www.azcentral.com/story/news/local/arizona-environment/2023/05/25/phoenix-is-not-prepared-for-a-simultaneous-heat-wave-and-blackout/70252691007/

24 For a comparable film, see the neo-noir science fiction movie "Dark City" (1998) at: www.imdb.com/title/tt0118929/

25 Eric Klinenberg, *Heat Wave: A Social Autopsy of Disaster in Chicago* (University of Chicago Press, 2002), 2015. See also "Dying Alone An Interview with Eric Klinenberg" at: https://press.uchicago.edu/Misc/Chicago/443213in.html

26 Lauren Sommer, "Why Heat Wave Warnings Are Falling Short in the U.S" *National Public Radio* (Sept. 13, 2022) at: www.npr.org/2022/09/13/1122491546/why-heat-wave-warnings-are-falling-short-in-the-u-s?utm_source=twitter.com&utm_medium=social&utm_campaign=npr&utm_term=nprnews

27 See Margaret Gulllette, *American Eldercide* (University of Chicago Press, in press).

28 Eric Klinenberg, *Palaces for the People: How Social Infrastructure Can Help Fight Inequality, Polarization, and the Decline of Civic Life* (Vintage, 2020).

29 World Health Organization, "Heat Waves" at: www.who.int/health-topics/heatwaves#tab=tab_1

30 Catherine Clifford, "Miami's First-Ever Chief Heat Officer on the Climate Issues That Scare Her the Most—and What Gives Her Hope" *CNBC* (July 9, 2021) at: www.cnbc.com/2021/07/09/miami-chief-heat-officer-shares-climate-issues-that-scare-her-the-most.html

31 Jack Healy, Edgar Sandoval and Elena Shao, "When the Heat Can't Be Beat" *NY Times* (July 28, 2022) at: www.nytimes.com/2022/07/28/us/heat-records-summer-climate.html?referringSource=articleShare

32 Maggie Koerth, "Why the Same Temperature Can Feel Different Somewhere Else" (Aug. 5, 2022), FiveThirtyEight at: https://fivethirtyeight.com/features/heat-index-temperature/?utm_source=pocket-newtab

33 Sanober Naheed and Salman Shooshtarian, "A Review of Cultural Background and Thermal Perceptions in Urban Environments" *ORCID Sustainability* 2021;13(16):9080 at: https://doi.org/10.3390/su13169080

34 Carbon Brief (Feb. 17, 2023) at: https://preview.mailerlite.io/emails/webview/249617/80364017821943310

35 "Heat Wave and Blackout Would Send Half of Phoenix to E.R., Study Says" at: www.nytimes.com/2023/05/23/climate/blackout-heat-wave-danger.html

36 www.worldweatherattribution.org/western-north-american-extreme-heat-virtually-impossible-without-human-caused-climate-change/

37 Heather Chen, "This Country's Love Affair with Air Conditioning Shows a Catch 22 of Climate Change" *CNN* (June 9, 2023) at: www.cnn.com/2023/06/09/asia/air-conditioning-singapore-climate-change-intl-hnk-dst/index.html

38 Coral Davenport, "Nations, Fighting Powerful Refrigerant That Warms Planet, Reach Landmark Deal" *New York Times* (Oct. 15, 2016).

39 Brookings, "As Extreme Heat Grips the Globe, Access to Air Conditioning Is an Urgent Public Health Issue" at: www.brookings.edu/blog/the-avenue/2022/07/25/as-extreme-heat-grips-the-globe-access-to-air-conditioning-is-an-urgent-public-health-issue/?utm_campaign=Brookings%20Brief&utm_medium=email&utm_content=220781934&utm_source=hs_email

40 Beth Daley, "Does Turning the Air Conditioning Off When You're Not Home Actually Save Energy? Three Engineers Run the Numbers" *The Conversation* (Aug. 22, 2022) at: https://theconversation.com/does-turning-the-air-conditioning-off-when-youre-not-home-actually-save-energy-three-engineers-run-the-numbers-188694

41 "Earth Talk: Will Air Conditioning Be the Death of Us?" *Tribune News Service* (*Environmental Magazine*, 2022) at: www.hastingstribune.com/ap/lifestyle/earth-talk-will-air-conditioning-be-the-death-of-us/article_6f03a3ed-d258–57ed-a638-574263ab3989.html

42 *ScienceDaily* (Feb. 4, 2022) at: www.sciencedaily.com/releases/2022/02/220204093124.htm

43 Adam Taylor, "Why European Homes (Usually) Don't Have Air Conditioning" *Washington Post* (July 20, 2022) at: www.washingtonpost.com/world/2022/07/20/europe-uk-air-conditioning-ac/?utm_source=newsletter&utm_medium=email&utm_campaign=wp_climate202&wpisrc=nl_climate202

　See also: Pranshu Verma, "Europe Is Overheating. This Climate-Friendly AC Could Help Key Point: Heat Pumps Are Efficient and Eco-Friendly. So Why Are They So Rarely Used?" *Washington Post* (July 21, 2022) at: www.washingtonpost.com/technology/2022/07/21/europe-heat-wave-heat-pump/?utm_source=newsletter&utm_medium=email&utm_campaign=wp_climate202&wpisrc=nl_climate202

44 Stan Cox, *Losing Our Cool: Uncomfortable Truths about Our Air-Conditioned World* (The New Press, 2012).

45 Eric Dean Wilson, *After Cooling: On Freon, Global Warming, and the Terrible Cost of Comfort* (Simon & Schuster, 2021).
46 Eric Dean Wilson, "Air Conditioning Will Not Save Us" *Time Magazine* (July 22, 2022).
47 Ibid.
48 Saugat Bolakhe, "Rethinking Air Conditioning Amid Climate Change" *Knowable Magazine* at: https://knowablemagazine.org/article/food-environment/2022/rethinking-air-conditioning-amid-climate-change?utm_source=pocket_collection_story
49 For alternatives to air conditioning, see: Matt Jancer, "How to Stay Cool Without Air-Conditioning" *WIRED Magazine* (Aug. 10, 2022); and Saugat Bolakhe, "Rethinking Air Conditioning Amid Climate Change" *Knowable Magazine* (Apr. 12, 2022).
50 *The Economist* (Nov. 11, 2021).
51 "Not Everyone Can Afford Air Conditioning During a Brutal Heat Wave. Here's How They Cope" at: www.cnn.com/2022/07/19/economy/air-conditioning-electric-bills-heat-wave/index.html?cid=external-feeds_iluminar_yahoo
52 Shannon Osaka, "Air Conditioning Has a Climate Problem. New Technology Could Help" *Washington Post* (Sept. 12, 2022) at: www.washingtonpost.com/climate-environment/2022/09/10/air-conditioner-ac-unit-climate-change/ See also "Harvard Scientists Develop Air Conditioning Alternative Much Better for Planet" at: www.msn.com/en-us/health/medical/harvard-scientists-develop-air-conditioning-alternative-much-better-for-planet/ar-AA113dPY?ocid=msedgdhp&pc=U531&cvid=fa8384532aa94de5ce6f43dc735135aa
53 "What's the Best Way to Fight Climate Change?" www.eib.org/en/surveys/climate-survey/3rd-climate-survey/best-ways-to-fight-climate-change.htm
54 Devon Thorsby, "6 Alternatives to Traditional Air Conditioning to Consider" *US News & World Report* (Aug. 19, 2022) at: https://realestate.usnews.com/real-estate/slideshows/6-alternatives-to-traditional-air-conditioning See also: "How to Stay Cool Without Air-Conditioning" *Wired* (June 20, 2023) at: www.wired.com/story/how-to-stay-cool-without-air-conditioning/
55 In some of the hottest places on earth, including Egypt and Turkey, there are traditional ways to keep cool. See: Philip Kennicott and Sima Diab, "Climate Solutions: Ancient Elements of Cool" *Washington Post* (Dec. 28, 2023).
56 Gulrez Shah Azhar, "Life Hacks from India on How to Stay Cool (Without an Air Conditioner)" *Goats and Soda: Stories of Life in a Changing World* (Aug. 2, 2022). See also "This Ancient AC System Will Cool Your House Without Electricity" at: www.fastcompany.com/90793729/this-ancient-ac-system-will-cool-your-house-without-electricity "How to cool your home without relying on air conditioning" at: www.washingtonpost.com/climate-solutions/2021/07/23/passive-cooling-heat-wave/?utm_source=pocket_collection_story.
57 "Passive Cooling" Could Reduce Indoor Temps by Up to 25 F in a Heat Wave" Researchers Have Found that Simple Passive Cooling Techniques Can Have a Surprisingly Large Effect at: www.freethink.com/environment/passive-cooling "How to Keep Cool When You Don't Have Air Conditioning at Home" at: www.ecowatch.com/staying-cool-tips-heat-climate-change.html
58 "50 Years after the First Earth Day, Environmental Activists Confront Another Global Crisis" at: www.azcentral.com/story/news/local/arizona-environment/2020/04/22/50th-anniversary-earth-day-another-crisis-looms/5157375002/

5   Drought
Where Did All the Water Go?

Drought is the last of the Four Horsemen of the Climate Apocalypse. It is listed last only because drought means sheer absence or the lack of water. Drought therefore stands in contrast to the first three of the Horsemen who directly invade our lives: fire, water, and heat wave. But the absence of water can also be devastating.

There are different definitions of drought, beginning with inadequate yearly precipitation in a specific region, insufficiency of water for agriculture, low water volume in streams or reservoirs, or drought that happens when demand exceeds the supply of water. By any definition, the Intergovernmental Panel on Climate Change foresees drought increasing in the southwestern USA, as well as other parts of the globe. More than half of the largest lakes and reservoirs in the world have been losing water, and in the USA we've seen the impact with dramatically lower water levels at the Hoover Dam and in Lake Mead.

One thing we can say for sure about water is that we take it for granted—that is, until we're thirsty. And just as death makes us appreciate life, so too drought makes us appreciate water, even if we don't understand much about it.[1]

One significant dimension of global warming is drought:

It is well established that parts of the planet are getting much drier. The Western U.S. is experiencing a period of dryness so extreme that some experts say it no longer qualifies as just a drought. Rather, it might be described as a more transformative trend of "aridification"—the most significant drought to descend on the American West in 1,200 years.[2]

Lower water levels in rivers and reservoirs are the most obvious sign of drought.[3] But there are also additional effects:

Drought and heat combined are putting the world's major staple crops—corn, wheat, soybeans, and rice—at risk, raising concerns about food insecurity in the absence of global leadership on climate change. Plummeting water levels in key reservoirs are also threatening hydroelectric

DOI: 10.4324/9781003345992-7

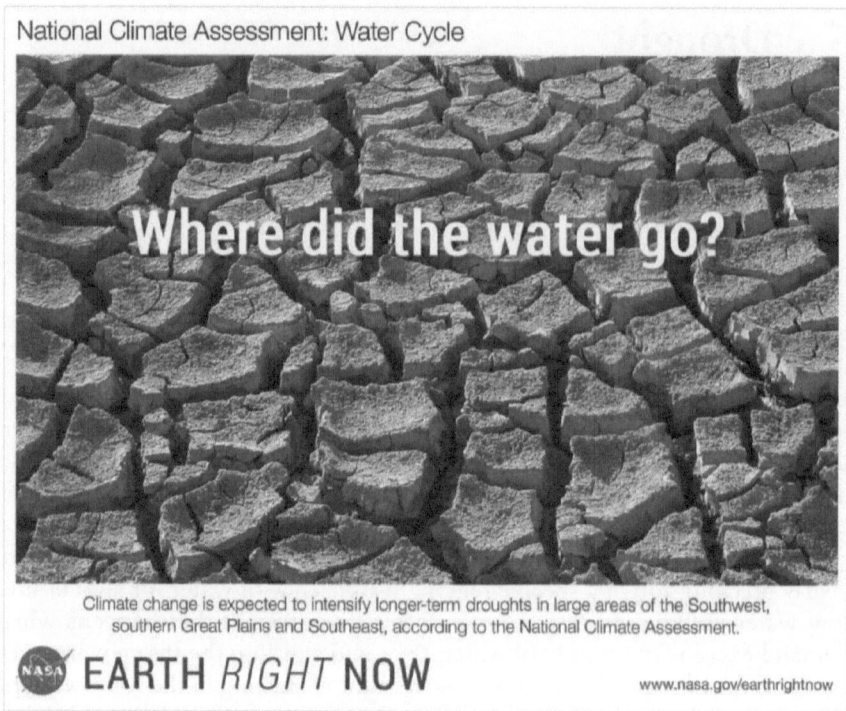

National Climate Assessment: Water Cycle

Where did the water go?

Climate change is expected to intensify longer-term droughts in large areas of the Southwest, southern Great Plains and Southeast, according to the National Climate Assessment.

EARTH *RIGHT* NOW   www.nasa.gov/earthrightnow

Figure 5.1 National Climate Assessment: Water Cycle. Public domain.

production, straining the electric grid and potentially leading to blackouts across the U.S. West. This is no longer a conversation about what could happen but a rallying cry for us to start exploring and elevating solutions.[4]

It is estimated that half of the world's population will experience water scarcity by 2025. As a growing global population and climate change have increased demand, water resources are being stretched further than ever to meet the needs of the people and industries that depend on them. Recycling and reusing water can combat water scarcity worldwide and provide communities, businesses, and local ecosystems with plentiful and secure resources. And the impact will be global.

The big impact of drought is the threat to our water supply, both for agriculture and for drinking water. By some estimates, up to half of the world's population could experience water scarcity in the years after 2025. With 8 billion people on the earth, climate change means increased demand for water at a time when weather patterns and precipitation are changing. For some time, farmers in the Midwest have been relying on groundwater from aquifers, and those aquifers are becoming used up, with less capacity to be

Figure 5.2 Barro Seco (Dry mud) in Chile. Public domain.

recharged during a period of drought. It means we are putting future generations at risk.[5] Recycling or reusing water could be an alternative.

Drought and Aging

Drought has special significance for an aging society. There is little doubt that drought is bad for human health, which means that it should be of concern for healthcare providers in an aging society.[6] A study from the National Institutes of Health, which tracked 11,000 older American adults over a period of 25 years, found that poor hydration was associated with high blood pressure, high cholesterol, high blood sugar, and even risk of premature death.[7]

Most studies of drought focus on its impact on agriculture or water systems more generally. But aging itself, in essence, means diminished reserve capacity. We need to look at the vulnerability of regions and also at vulnerability when it comes to individual health with aging.[8] The bottom line is that the outcomes of drought are difficult to assess just as the risk assessment itself is difficult, even when we limit ourselves to the lower 48 states of the USA.[9]

Lessons From History

The most widespread of all natural disasters is drought, as we see through history. But drought is different from other natural disasters because it isn't an event. It is rather a non-event: a time when something doesn't happen, a time when it doesn't rain.

History has lessons for us about drought and other environmental threats.[10] Eric Cline, in his masterpiece *1177 BC: The Year Civilization Collapsed*, analyzed the series of events that led to the collective collapse of multiple Bronze Age societies, including Minoans, Mycenaeans, Babylonians, Hittites, and Egyptians—all powerful centers of civilization, all going down within a century. Like other historians, Cline points to climate change—in particular, drought—as one of the major causes (but not the only one) in a cascading series of events. These events led to the "domino effect" among the societies connected by trade routes, as if by "globalization." It is a remarkable parallel with our time, obviously.

The story of this Bronze Age collapse underscores the importance of drought as a factor in history, perhaps in parallel to what historian William McNeill showed about disease in *Plagues and Peoples*. We can identify at least ten civilizations or nations that collapsed from drought, and so, as Jeff Masters puts it, "We should not grow overconfident that our current global civilization is immune" from the threat of drought.[11] Historians tell us how people abandoned prosperous cities in Mesopotamia, the Indus Valley, and beyond when a decades-long drought threatened them and others around the world. One example is shown here: the ruins of a Middle Eastern civilization that disappeared because of drought. In Nabatean texts in Greek, *Ba'alshamin* is identified with Zeus Helios, or Zeus as the sun god. When drought happened, the inhabitants stretched out their hands praying for rain. But the rain didn't come.

There's a big difference between drought and the other Three Horsemen of the climate apocalypse discussed here. Fire, flood, and heat waves are mostly limited in time: these disasters can each last for a few weeks, though they may repeat, of course. Drought typically lasts much longer. A drought like the Dust Bowl might last up to a decade, and a megadrought could last for decades or more, as the drought in California threatened to be before it ended in 2023.[12] Drought could well be a major element in the collapse of the Bronze Age societies, just as it probably was for pre-industrial societies such as the Anasazi of the American Southwest, the Khmer Empire of Cambodia, the Mayas of Central America, and the Yuan Dynasty of China.[13]

Climate scientists believe that the current southwestern North American megadrought was the driest period of two decades in this area since around the year 800 CE.[14] By 2022, the diminished water supplies began to seriously affect vegetable prices in California, as I learned from personal experience and as most Americans learned in coping with unexpected inflation.

We have been looking at climate risk in terms of disasters like forest fires, hurricanes, and heat waves, which all have a disproportionate impact on older people. But drought is the most widespread natural disaster, and evidence shows that it can be just as fatal to older people. A study of health conditions over a period of 14 years found that severe drought conditions increased the death rate for adults 65 or over. The mortality rise was found even in places far removed from southwestern states familiar with drought. For example, counties in Minnesota displayed a higher risk of mortality and cardiovascular disease than would be expected. Jesse Berman, the lead author on this research at Yale, emphasized the importance of these findings because climate change is anticipated to increase the frequency and severity of droughts.[15]

Drought and the USA

Water shortages are the most immediate sign we know that drought is here. On a global basis, people are facing a condition where they have to live with less water. Global warming has begun melting glaciers and thereby reducing snowpack, which is the source of water in many parts of the USA, especially the Southwest where I have lived. For example, Los Angeles and its suburbs may soon be putting limits on outdoor water. Water management officials expected drought but not this extreme and not this fast.[16]

The area of Rio Verde Foothills in Arizona is near Phoenix, and it has gotten its water from Scottsdale, which in turn relies for the most part on the shrinking Colorado River. Developers build houses in Rio Verde, and now residents do not have access to water.[17] The Governor of Arizona in 1012 announced that the state would no longer permit development in the Phoenix area relying on groundwater, a decision that would force developers and cities to turn to other, more costly sources.[18]

As has been widely reported, California, the state where I live, endured a massive megadrought. In May 2022 federal officials took unprecedented steps in response to declining water levels of the Colorado River, which is a key source of water for many in the Southwest USA. Water in the Colorado River is also a key source of electricity. Officials were able to use various "workarounds" to keep some reservoir levels high enough to avoid loss of electricity to the millions of people who depend on it. By the end of 2022 conditions on the Colorado River had only gotten worse, and the future of conditions under a megadrought remains unknown. For the millions of people who depend on Glen Canyon Dam for electricity, drinking water, and irrigation of crops, two decades of drought have brought a reckoning.[19]

Public discussion about drought revolves around the Southwest USA. How bad is that current drought? How about the worst in 1,200 years?[20] And climate change is a major factor. Think about that time scale: as far back as the time of Charlemagne (800 CE). Climate change, and drought, is not just the future. It's here and now.

During the summer of 2022, 6 million people in California faced unprecedented water restrictions at a time when the state was coping with the worst drought in recorded history. Both residents and businesses in Southern California counties, including Los Angeles, were required to limit outdoor water use to a day or two per week. All water users were urged to cut water use by roughly 25%. In Texas, a severe drought was threatening the state's water supply, not to mention farmers who depend on water for their crops. By the end of 2023, the California drought ended. But drought still remains an unavoidable part of California's climate. In the past 40 years few large-scale projects have been built, despite the doubling of the population.

The year 2022 was a major year for drought, but, for California and other parts of the western USA, it was just the latest in years of a megadrought, one of the worst in many decades.[21] Ironically, the signs were evident nearly a decade earlier.[22]

We need to notice things where "nothing happens," and we need to notice places beyond the Southwest. So-called flash droughts—comparable to flash floods—have now become a reality. Federal drought maps show that parts of New England are in "extreme drought." Mid-Atlantic states also are shown to be in "severe drought." I spent much of my own earlier life in New York

Drought conditions at the end of April 2024

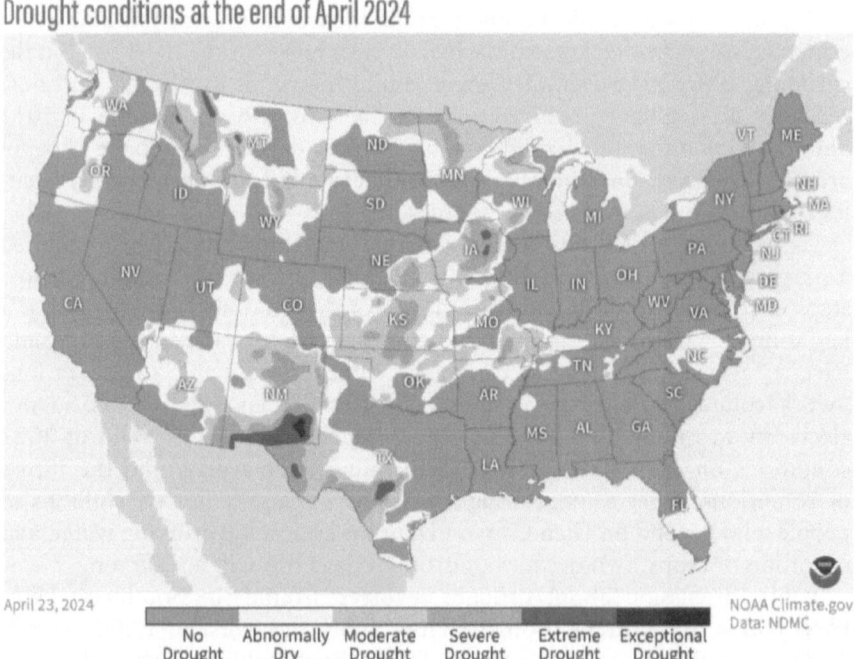

Figure 5.3 U.S. Climate Outlook (May 2024). NOAA Climate.gov map based on U.S. Drought Monitor Project data.

State, where drought was simply never part of the world I grew up in. Water was just what came out when you turned on the faucet. The world we elders are leaving to younger generations is different. The "new normal" of our time now includes the existence of megadroughts.[23]

Drought, like others of the Four Horsemen of climate apocalypse, varies in different parts of the USA. Since I live in California, I have looked in particular at the region encompassing Arizona, California, Colorado, New Mexico, Nevada, and Utah. In my area, drought is clearly a major threat:[24]

> The hottest and driest corner of the country is already suffering from heat waves, droughts, and increased wildfires. As a result, the Southwest, to put it bluntly, is running out of water. With water at already record low levels and a population that continues to grow, the region is working on a recipe for water scarcity.

Many details about different regions of the USA are available Fourth National Climate Assessment (2022). Patrick Gonzalez, writing in the National Climate Assessment, looks closely at the Colorado River and Lake Mead:

> Lake Mead, which provides drinking water to Las Vegas and water for agriculture in the region, has fallen to its lowest level since the filling of the reservoir in 1936 and lost 60 percent of its volume.

Clearly, the Southwest will be a major site for drought in years to come.[25] Elizabeth Kolbert[26] in her article "Climate Change from A to Z" concludes her account with a visit to Hoover Dam, located between Nevada and Arizona, and a major path for the Colorado River.

"The Colorado River basin has been called ground zero for climate change in the United States." If this is the case, then Hoover Dam might be described as ground zero's ground zero. Since 1998, the basin has been stuck in a drought; this drought has lasted so long and grown so deep that it's now routinely referred to as a megadrought.

U.S. House Speaker Tip O'Neill once said that all politics is local. The same could be said about drought and many other extreme weather events. But, in the case of drought, "local" has gotten much bigger because of global climate change. Hot and dry conditions now pervade many Western States, including the seven states in which the Colorado River runs: Colorado, Wyoming, Utah, New Mexico, California, Arizona, and Nevada.

Drought is connected with diet as a key element in climate change. For example, the Colorado River, a major source of water for Southwestern states, has been faced with the impact of drought. But what is all that water used for?[27] The answer is that most of the water—we're talking about nearly 80%—is used for agriculture, and much of that goes to grow feed for livestock. In short, drought is tied directly to diet, to how much meat and dairy we are eating. Some of the water goes for other agriculture, like vegetable farming. But it's less than a quarter of the total amount of water that goes for livestock.[28]

This pattern of water usage means that, however well intentioned, we will not solve the water crisis by taking shorter showers or by not watering the lawn. For comparative purposes, it would take 38 gallons of water to get a quarter-pound beef patty. Why so much? Because we have to include all the water needed to grow the feed for cattle. By contrast, we could get the same amount of protein from a vegetable source like tofu, with only five gallons of water. Dairy products, like milk and cheese, are even worse at consuming water. Instead of speaking about a "carbon footprint," we can speak about a "water footprint" and the bottom line is clear. Beef and dairy are intensely water-consuming foods, even without considering methane produced as a harmful greenhouse gas from raising cattle.

What drought in the Colorado River is showing us is that diet and an unsustainable lifestyle is the driving force behind the problem. Instead of praying for rain, we need to change what's on the dinner table, discussed in more detail in the chapter on "Becoming a Climate-Conscious Consumer."

The Colorado River, now in drought condition, is a problem that's been building for decades. It did not disappear because of unusual rainfall in January 2023. But drought means hard decisions:

> How do you balance securing water for some of the fastest growing cities in the country with preserving water for farming regions that supply most of the country's winter vegetables? Do you recognize that some users have a longer established right to the water than others, such as farming regions that laid legal claim to the water before urban areas?[29]

By 2023, heavy snowfall has helped replenish Lake Mead and other reservoirs. Snowmelt after a wet winter in the Colorado River basin was beneficial for Mead. A quick fix, from unexpectedly welcome rainfall along with negotiated cuts, has bought some time.[30] The question will be whether that snowmelt will impact the prospect of rationing limited Colorado River water. States like California and Arizona are directly affected by long-term drought that continues. As always, the outcome is uncertain.

The drought, or megadrought, has already had a dramatic impact on the hydroelectric turbines that generate electricity at Hoover Dam. Because of the drought and water shortage in Lake Mead, Hoover Dam's 17 turbines now operate only sporadically, thus generating less electricity. Here is the irony. Electrification is understood to be a major power alternative to fossil fuels, and hydroelectric power is a key element in generating electricity. But drought, in the southwest as in other regions, is cutting back on the capacity to produce electricity at all.

Future Prospects

What is the future of drought in the USA?[31] Predictions are uncertain, but scientists believe that we may be in for years of higher temperatures and diminished rainfall. The big challenge is that people today, including older Americans, are moving to some of the hottest, driest, and most vulnerable parts of the country. In short, they are moving where the water isn't—to Sunbelt states where the housing is less expensive, but the risk of drought and climate change is getting worse.[32] The Sunbelt was once the retirement destination of choice, as it was for my own parents, moving from Michigan to Florida. That was before climate change.[33]

Climate-induced megadrought threatens the future of the Southwest USA. In light of it, we cannot avoid questions about unrestrained economic growth. Will there be enough water for people in the future? In Nevada and other states, developers in recent decades have built millions of homes, despite relying on shrinking water sources. Real estate companies and home building companies have based their financial hopes in the Southwest even as the region continues to run out of water. Some major companies, like D.R. Horton, have gone so far as to buy up water companies to be sure of having that precious resource for their continued growth and expansion.[34]

Drought is clearly one of the big threats in a warming world. But we should be aware of falling into a false dichotomy: drought is a lack of water and is the opposite of flooding. But drought also involves a paradox:[35] climate change with drought clearly brings about a hotter and drier world, more prone to wildfires. But climate change also promotes the circulation of moisture into atmospheric rivers where storms are supercharged and dangerous, as residents in California learned in the first two weeks of 2023. In many parts of California, both cities and rural areas, extraordinary rainfall meant flooding, including the loss of life. Property damage was on the scale of hurricanes in other parts of the country. In California, areas stripped of trees could not absorb water and were therefore subject to mudslides. The earlier discussion of fire and flood is thus linked to the threats posed by drought as well. In short, we need to avoid the false dichotomy of drought versus flood. Both can happen, in the same place, one right after the other.

I encountered these contradictory facts about flooding from personal experience. On the very last day of 2022, New Year's Eve, my wife and I unexpectedly ended up driving through an atmospheric river that engulfed the San Francisco Bay Area, drenching everything with the second-worst rainstorm in 170 years. The floods have continued for the following two weeks, as well. As we drove through streets and encountered floods, barely able to drive home, I wondered: Can this be a bad dream? Such unexpected flooding is likely to be more frequent in years to come. In retirement, we live at the top of high hills on the Bay Area Peninsula. We were not searching for safety in old age. We thought we had found it. But we were wrong. It was not a bad dream but reality, the "new normal" in which we are living.

Can we rely on past weather experiences to predict the future? I thought of the report titled, "Death Valley Experiences 1,000 Year Rain Event" about the devastating floods in Death Valley, California.[36] These were floods which stranded a thousand people. In 2015, my wife and I were on a Road Scholar trip to Death Valley. But some attractions, like Scotty's Castle, were closed to visitors because of the floods that came that year: less than a decade and a lot less than every thousand years.

This discussion has focused on the Southwest USA. But the threat of drought is not limited to that part of America. If we look at a map, the 100th meridian marks the division between the dry West and the more humid East. That marker is now slowly moving eastward because of climate change. What will be the impact of that shift on food systems and water shortages? Even those who are now older people are likely to find out sooner than they expected. The meridian marker has already moved about 140 miles since 1980, according to climate scientists. Richard Seager led two major studies predicting that, if this drying trend goes on, farms further to the east will be directly threatened. Adaptation means irrigation, or, without that, changing crops. But the Union of Concerned Scientists, among others, has warned that farm size and crop consolidation (monoculture) may demand more fertilizers and pesticides. The threat is that large croplands could fail.[37] The threat of drought is here and now.

Where Is the Hope?

Are there solutions to the challenge of drought? Yes, rationing water is one of the solutions, but it can have devastating consequences: for agriculture and, given the importance of hydration, direct consequences for human health. Robert Costanza, an ecological economist at the Institute for Global Prosperity at University College London, understands that small incremental changes may not solve the problem of drought. A deeper approach, sometimes called bioregionalism, is called for.[38] In Robert Costanza's words:

> A lot of the things that we're talking about with the circular economy—regenerating wetlands, planting forests, dealing with climate change—are difficult to implement because the underlying goal is still GDP growth, and these things get in the way of that.[39]

Just as we talk about "slow food" or "slow money," so too we will need to talk about "slow water."

Without more changes to water management, planet Earth could face a 40% shortage in water supply by 2030, according to a 2016 report by the International Resource Panel of Scientific Experts. One of the solutions recommended is to store water the way that nature does, as some in California are recommending. Increasing water supply will also involve diminishing and decreasing pollution, eliminating dumping, avoiding dangerous chemicals, cutting in half the share of untreated wastewater, and increasing recycling, where that is possible.

On a personal level, I have very little experience of drought. I turn on the faucet every day, and, like most people, I take water for granted. But, in two years, 2022 and 2023, California went from drought and threats of water rationing over to record-breaking snowfall and flooding. This back-and-forth between drought and flooding may become the "new normal" in a time of climate change. The San Francisco Public Utility Commission, like others in California and throughout the Southwest, will do its best to maximize storage capacity, but, as Christopher Graham, manager of the Utility Commission said, "climate change is going to make managing this water supply much more difficult."[40]

SEARCHING FOR SAFETY

I was born in the last year of WWII in Boca Raton, Florida, on the east coast, in the path of future hurricanes. Soon after my mother took me to Chicago, the site of the devastating 1995 heat wave, and then later to Merrick, Long Island, the target of Hurricane Sandy in 2012 (I had left long before). My wife and I owned a time-share on Hilton Head Island in South Carolina: on the ground floor, right next to the beach. Learning more about climate change, we eventually sold it. In retirement we later moved to Boulder, Colorado, living right on open space—comfortable until my neighbors told me about when they were evacuated because of fire threat. Before I moved again, evacuation wasn't needed, but I had close friends who were evacuated during the vast Marshall Fire of 2021 outside Boulder. That fire burned over 6,000 acres and did damage estimated at half a billion dollars.

By the time of the Marshall Fire, we had already moved to California. Just days after we arrived, midday smoke began to cover the neighborhood. Our granddaughter noticed it was only one of many wildfires in the State, including one in Edgewood Park in 2022 that cut off our own electricity for a time. When I look back at my own life, I began to wonder: Where do we go in searching for safety from extreme weather events prompted by climate change?

We ended up living in an apartment a few miles from the San Andreas Fault, the place where I'm writing the words you're reading now. If not climate change, then earthquakes? Again: Where do we go in searching for safety?

HERE—NOW—YOU—HOPE

Here's a message from my hometown, New York City, where we raised our children and where I spent much of my work life. David Wallace-Wells said when smoke from wildfires in Canada came over New York:

> Until now, if people in the green and leafy Northeast looked at arid Western cities covered in smoke from wildfires, they could say, that

can't happen here, thank God. On Tuesday, it did: For a moment, New York's air quality was worse than it was in Delhi, the infamous pollution capital where average life spans are reduced more than nine years by particulates in the air[41]

Have we forgotten the Arabic fable "Appointment in Samarra?" As that story goes, one day the Angel of Death accidentally meets a man in Baghdad and the Angel is startled at seeing him. The man is terrified by the Angel of Death and he frantically runs off to the city of Samarra. When they ask the Angel of Death why he reacted with surprise, the Angel replies, "I was astonished to see that man in Baghdad because I have an appointment with him tonight in Samarra." Again, where do we go in searching for safety?

Growing older in the 21st century presents a big challenge. Heading for the sunbelt makes less sense if you're heading into blistering heat waves, powerful storms, frequent floods, and dangerous wildfires:[42]

For older Americans, especially, climate change can take a significant toll on health. Carbon dioxide emissions, a major contributor to climate change, can directly aggravate heart disease, lung conditions and diabetes, for example, while climate change itself increases ground-level ozone, dust from droughts and smoke from wildfires, all of which contribute to breathing problems . . .

A recent poll shows one in five people are so concerned about global warming they believe it will make it more challenging to live in their area. And data show relocations due to climate change are already happening. In the U.S., alone, wildfires, storms, floods and other disasters displaced 543,000 people in 2022, according to the Internal Displacement Monitoring Centre . . .

Ideally, people will listen to the science around climate change when making their housing decisions, [NASA climate scientist Sharon] Miner adds. "I'm hoping that people start to consider climate change in some of their planning and their choices . . . I do think that it's increasingly needed."

Climate scientists estimate that extreme wildfires may increase by 14% globally by 2030, despite efforts to curb greenhouse gas emissions, according to the United Nations Environment Program (Brown, 2022). The U.N. report has noted a "dramatic shift" in wildfire patterns because of global warming and patterns of land use. Just as I did when in retirement I moved right next to the open space in Boulder, Colorado, more and more people are choosing to live in proximity to places where wildfire will be in their future.

Today an increasing number of older people are considering climate when thinking about where to live in retirement. Real estate agents understand that retirees will be looking beyond taxes or crime rates, and they are beginning to think about extreme weather events, including wildfires and floods.

A study of nearly 1.4 million home sales in coastal Florida revealed that between 2013 and 2018 the sales volume for properties declined by a fifth. The impact of Hurricane Ian in Florida in the fall of 2022 would likely show a similar decline since two-thirds of deaths from that hurricane were people over the age of 60. Home prices in riskier areas in Florida declined between 2018 and 2020.[43]

Two major trends will already be coming together in the 21st century: the demography of population aging and the force of climate change. Disasters of fire and flood will have an overwhelming impact on an aging population because that population is aging in place in locations vulnerable to these threats.

For those who are looking for a place to live in retirement, there now are tools available to help identify risks: for example, the National Oceanic and Atmospheric Association has a website, Climate.gov, which identifies data about floods, wildfires, drought, and similar other climate hazards. Another website, Risk Factor, was developed by the nonprofit First Street Foundation. Risk Factor draws on multiple data sources to quantify climate risk for home buyers.

Older people who are simply aging in place out of inertia or a lifelong pattern may be unlikely to consult such sources. But knowing about climate risk could guide adaptation steps, such as protecting a house from flood and fire or planning for insurance for risks. It's another matter to think about finding another place to live, either before or after a disaster strikes. Planning for retirement means searching for safety, even finding air to breathe safely.[44] We return to that point later in examining what searching for safety means in an era of climate change.

The point was stated well by Dr. Ed Kearns, chief data officer of First Street Foundation: "Very often, you hear people say, 'I don't have to worry about climate change because I am 60 or 70 years old—this is a problem for my grandkids,'" he said. "I greatly disagree with that statement. Climate change is here, and it is affecting all of us."[45] I thought of the older woman who listened to my climate speech and said "I'm glad I won't live to see this."

There is a Sufi story about a king fearful of threats that could harm him. So the king withdrew to a castle surrounded by a moat. Then inside the castle, he became worried, so he moved into a single room far from potential harm. But the room had windows, so he sealed them up and then replaced the door with a wall of bricks, leaving only a single opening. Then the king came to feel that this opening, too, was a place where harm could get to him, so he sealed that up, too. And that is how the king died. So goes the Sufi story.

The condition of the king is our condition with respect to climate risk and aging in place. We're searching for safety, but climate change poses threats on every side: What can be called the Four Horsemen of Climate Apocalypse— wildfire, flood, drought, and heat waves. What to do?

Every survey undertaken shows that older people, people like me, as a group, will overwhelmingly say that they want to age in place. Perhaps they mean only that they don't want to go into a long-term care setting. But their behavior suggests that, as a group, seniors are reluctant to move. True, not everyone: I myself have moved four times since I turned 65, with good reasons in every case. It's natural to create a home that feels comfortable and familiar and only natural to want to stay there.[46] Proponents of aging in place can tell us things to make our home safer—whether grab bars in the bathroom, lights on the stairs, or emergency response capacity ("Help, I've fallen"). With advancing age, there are more threats that harm us. Like the king in the Sufi story, we're searching for safety, and climate change is a threat. Those tempted by a head-in-the-sand posture can try to ignore the facts. But reality has a way of imposing itself, and we all do better by facing the facts upfront. James Baldwin put it best: "Not everything that is faced can be changed, but nothing can be changed until it is faced."

In these first two chapters, we explored two kinds of climate risk that are present and likely to become worse in years to come: wildfire and flood. These are the first Two Horsemen of the climate apocalypse and they both threaten to destroy individual homes. True, not all places in the USA are equally susceptible to each of these threats. Florida, for example, is uniquely vulnerable to hurricanes, but so are those in many states on the East Coast. California is vulnerable to wildfires, but so are many parts of the Western USA.

California is the nation's biggest insurance market, and what happens there has implications for U.S. consumers,[47] as carriers leave disaster-prone states like Florida or California. California wants to avoid becoming like Florida, where homeowner premiums rapidly skyrocketed. Florida residents pay $6,000 a year on average whereas Californians pay about around $1,300. But the future may not be like the past. Industry groups argue that Californians should actually be paying more considering exposure to major disasters as well as the higher price of homes.

In earlier chapters, we've looked more closely at climate risk from the other Two Horsemen of the climate apocalypse—heat waves and drought. Once again, the threats vary depending on location. Even more threatening, drought and heat waves threaten us even if they leave our homes undamaged. And it is not easy to predict exactly where these threats will arise. Former House Speaker Tip O'Neill famously said, "All politics is local." We can say the same thing about climate risk: every risk from the Climate Apocalypse is local. Planning and adaptation must be local, too.

How do we escape from fire, flood, drought, and heat waves? What will come from searching for safety in an era of climate crisis?

The answer forces us to face reality, as George Monbiot asks us to do:

> There are no perfect solutions in an imperfect world. Everything we might propose, including all the ways forward I suggest in [my book] *Regenesis*, has downsides. We are working in a very tight space, one in which 8 billion people and more need to be fed, within an earth system whose planetary boundaries have already been breached, to a large extent as a result of food production.

What is clear enough is that aging in place cannot be successful as long as we think of ourselves as living in a dwelling protected from harm. Searching for safety cannot be limited to making changes inside a home as if we remain vulnerable to fire and flood, drought, and heat waves. We may need to think about climate risk and aging in place differently.

Kathleen Sullivan was born in the same year as I was, 1945, and lived not far from me on the south shore of Long Island, although we have never met. But her words about climate change carry a generational message for this historical moment:

> The narrative I grew up with was that America was a great and valorous place where dreams could be realized and ambitions for material wealth and status satisfied in larger ways by each succeeding generation. There were no stories back then to temper this belief in our unending entitlement to expansion by gobbling up all the resources of the natural world.
>
> Nature, it turns out, is not too big for humans to have, in the course of a lifetime, created planetary-scale disruptions.
>
> As have many of my illusions formed as a child about who I am, who mankind is on the earth and what the earth needs from us in order to thrive. There is terrible grief in all this loss. And humility. I can choose among a glittering array of defense mechanisms for this pain: I can deny the dire emergency of this moment in time; I can dissociate and say it's not my problem, I'm too old, leave it to the young."[48]

What is called "climate grief" takes a particular form depending on our age. For younger people, it is the loss of a livable future. For older people it is the grief of lost illusions, as Kathleen Sullivan says so well. Whatever our place in the life course, we must strenuously resist that "glittering array of defense mechanisms" around the climate threat we all face. Resistance means avoiding false hope, but, above all, it means avoiding complacency and passivity. Genuine hope means action, which is only effective when based on reality.

5.1   Pete Seeger: Eco-Elder

Figure 5.4 Pete Seeger. Public domain.

Pete Seeger is well described as the father of the folk music revival, continuing the tradition of Woodie Gutherie and others. He was a prolific songwriter whose work includes "Where Have All the Flowers Gone?" and other songs known to millions. Pete Seeger was also a lifelong environmentalist who in 1966 founded The Hudson River Sloop Clearwater to protect the Hudson River. Seeger refused to accept the idea that the Hudson was hopelessly polluted, and so his efforts resulted in dramatic improvement of the river.

Pete Seeger's life and work had special meaning for me because I lived and raised our two children on the Hudson River. Our daughter spent time on Pete Seeger's Clearwater Sloop while my wife and I were able to attend the music and environmental festival, the Great Hudson River Revival. In later life, Pete Seeger referred to himself as an oldster and he continued performing for school children, who he said gave him hope for the future.

Pete Seeger died in 2014 at the age of 94, vigorous as an Eco-Elder until the end of his life.[49]

As seen in the preceding chapters, climate change is already a risk for aging in place. Extreme weather disasters in the USA affected more than 14 million homes in 2021, causing nearly $57 billion in damages.[50]

For those who are not in denial ("It won't happen") or not in complacency ("We'll find a way to cope"), there remains the challenge: How can we cope with the threats of the Four Horsemen of the Climate Apocalypse—fire, flood, heat wave, and drought? One answer is resilience and adaptation.[51]

For those looking for safeguards, there are two simple responses: get out of the way of the threat or get insurance to cover the threat that is coming. Getting insurance is familiar: life insurance, disability insurance, unemployment insurance, and so on. For climate change the most obvious example is homeowner's insurance, often available for fire or flood. But there is no comparable kind of insurance against the impact of drought or heat waves.[52] One answer in searching for safety is to get out of the way and move to another part of the country. Here we examine what searching for safety might mean.

In an earlier chapter, we looked at insurance as a strategy. For those who can afford it, there is insurance for fire or flood. Insurance is far less available for heat waves or drought. Another answer, of course, is emergency preparedness or adaptation in advance, which is why I have a "Go Bag" since I live in an earthquake zone.[53] Protecting your home, and your wallet, against natural disasters is a good idea, whether from earthquakes or extreme weather events.[54] The Environmental Protection Agency offers a climate change mapping instrument that covers floods, wildfires, and even drought.[55]

With growing fears about climate change, a small but growing number of people in the USA are already moving to places they consider to be safe locations to live, such as New England or the Appalachian Mountain states.[56] "Millions and likely tens of millions of Americans" will move because of climate through the end of the century, said Jesse Keenan, an associate professor of real estate at the Tulane School of Architecture.[57]

More and more people have seen the risks, including the risk of aging in place, and have decided to relocate.[58] Climate change in an aging society means facing rising housing costs, which push people into parts of the country more vulnerable to climate change. Retirement is one point where these housing choices are made.[59] "Over the past 20 years, counties and neighborhoods at high risk of extreme heat, droughts, wildfires, and floods have seen notably faster population growth than low-risk locations."[60]

Along with voluntary relocation, others today are raising the prospect of compulsory migration or what can be called "managed retreat."[61] A key point is that migration is not just a hypothetical scenario for the future. It is already happening within the USA.[62] Nearly a third of Americans pointed to climate change as a reason to move in 2022. A new survey conducted by *Forbes* found that almost a third of survey respondents cited worsening weather conditions as a reason to move, and over half of respondents who moved within the last few years reported their move to be unexpected.[63]

And it's happening on the Southern U.S. Border because immigration, legal or not, is driven by climate change in developing countries of the Global South. Rising sea levels prompt a U.S. migration crisis both within and across our borders.[64]

I saw the prospect of managed retreat firsthand in California. In Isla Vista, at Santa Barbara, California, where I teach in the aging program at Fielding Graduate University, there are apartment buildings that periodically lose their front patios or their facades to the rising ocean as it erodes the cliffs where the buildings stand. In October 2022, I visited Santa Cruz, California, where an exciting surfing competition was taking place. Along West Cliff Drive I knew that I was near the point where roads and homes were threatened by similar erosion from rising sea levels and storms. The California Coastal Commission has been discussing the threat with local officials in Santa Cruz. By the end of this century, property on West Cliff Drive could be facing up to a billion dollars' worth of damage from the ocean.[65] Our own family vacation spot located in Aptos is not far from Capitola, where storm damage in January 2023 was so severe that President Joe Biden paid an official visit. No wonder the California Coastal Commission has begun talking about managed retreats in a period of climate change.

Whether as a retirement dream or in searching for safety, people in later life do move and hope to avoid extreme weather events. But climate change is casting a shadow on such decisions.[66] A good example is the story of Bill and Annemarie Kachur, who retired in 2016, and saw no reason to leave their home in Myrtle Beach, South Carolina. Since then, following four hurricanes and two evacuations to higher ground, things look different. Aging in place is no longer so attractive: "The sixty-five-year-old is thinking about the seventy-five-year-old in me and do I want to be doing that in ten years?" says Bill Kachur. "And the answer is no." A survey from the real estate firm Redfin found that half of those who plan to move in the next year reported that heat waves or other natural disasters were a factor in the decision to relocate. More than a third cited rising sea levels.[67]

The issue of migration, aging, and climate change is illustrated by the case of Connie and Glenn Faast, who left their home in the Florida Keys for the mountains of North Carolina after nearly 50 years living in the Keys. "It was pretty much immediate," said Connie Faast. "It's just too hard to start over when you get older. We couldn't risk it."[68] In April 2023, ABC News reported that members of the Isle de Jean Charles Choctaw Nation and the United Houma Nation tribes had become the first communities in the USA to be displaced by climate change. They will not be the last.[69]

Breaking Our Hearts

Searching for safety is a flight from areas vulnerable to extreme weather events. Likely locations for floods or wildfires can be found on maps or other guidance for climate-ready communities.[70] Other maps can tell us about heat waves, such as the federal website heat.gov. In the case of heat waves, too, we may want to flee and search for safety:

Consider the story of Anne Morollo, age 57, from Phoenix, Arizona, who thought she could escape the terrible heat of Florida summer by moving to the foothills of the Rocky Mountains. After her first wildfire at her new location, that illusion of safety was shattered. Anne Morollo moved back South, to Savanna, Georgia. Today she is more or less homebound, far younger than one imagines for the very old:

> All she does is stay home, avoid the front seat of her broiling Volvo, and sink into air-conditioned isolation. "I always thought I could wriggle my way around this," she said. "But what can I do?"

Or consider 70-year-old Paolo Pinto of Austin, Texas, who described his situation coping with heat:

> After 10 o'clock, I'm inside the house . . . I have curtains, shades and fans. I don't come out until around 7 p.m. I turn red, I get exhausted.[71]

These stories tell us only about the logistics, about the coming and going, of searching for safety. The lived experience is something else, conveyed in the words of poet Mary Oliver:

> I tell you this
> to break your heart
> by which I mean only
> that it break open and never close again
> to the rest of the world.

Rachel Alexander[72] elaborates on Mary Oliver's message in these terms:

> "Here, Oliver acknowledges that the abundance of suffering in the world can lead many of us to narrow the scope of our care to just a few people. For some, the nucleus of this care is their families. The wind outside may howl, we think, but if we can build our little house and lock the doors tightly and huddle closely together, we'll be OK.

Central to this line of thinking is the fallacy of control: that our actions—including stockpiling resources for only our loved ones—can shield us from harm. Climate change is one thing that upends this belief. While those without money and power will suffer more than those who have them, the truth is that nobody is really safe.

Indeed, this is the terror of climate change. But it may also be its promise, ushering people like me into the sense of existential vulnerability that many have known for a long time. This vulnerability demands that we expand our circle of empathy—re-evaluate who merits our love and, in turn, our care."

Rachel Alexander concludes:

So my family will follow (Mary) Oliver's advice: We'll witness the beauty of what we have while it's ours, and act to protect and repair where it's possible to do so. We'll also choose to look at loss head-on, because telling the truth, even when it's scary, is part of loving, too.

Beyond vulnerability, Mary Oliver's poem suggests that we might need to question the basic premise—searching for safety. Perhaps the better approach is searching for solidarity. Klaus Jacob, a Columbia University climate scientist, believes that, in this climate crisis, we are in long-term denial, a pattern which is unsustainable and is putting future generations at risk.[73]

At the end of the day, how do we escape from fire, flood, drought, and heat wave? What is the result of searching for safety in an era of climate change? The answer demands that we face reality:

There are no perfect solutions in an imperfect world. Everything we might propose, including all the ways forward I suggest in *Regenesis*, has downsides. We are working in a very tight space, one in which 8 billion people and more need to be fed, within an Earth system whose planetary boundaries have already been breached, to a large extent as a result of food production.

There are no remaining comfort zones. There is no longer—if there ever was—scope for ideological congruence, for solutions that fit snugly into any one worldview. We will find ourselves in disconcerting places. We will be assailed by cognitive dissonance.[74]

Managed Retreat?

In this discussion we've looked at safety as a matter of individual choice. But as we think about responses to extreme weather events, at some point we have to think about decisions made for others, especially as a matter of public policy: what has been called "managed retreat."[75] Managed retreat means simply moving people or buildings out of locations repeatedly exposed to hazards like fire or flood. In a time of climate change, managed retreat means responding to predictable disasters such as sea level rise. Individuals may decide on retreat, as we've seen. In the case of aging, retreat may seem impossible because "I'm too old to move." But the opposite can also be true: "I'm too old to risk going through this again."

Whatever decisions individuals make, the government, at the local, state, or federal level, can act on managed retreat by buying the homes of people living in places vulnerable to the climate threat. Apart from protecting financial assets, areas where people leave can be restored to natural habitats, as in the case of New Orleans, where wetlands are a significant buffer against hurricanes and floods.

Managed retreat can be a first choice, or it can be a last resort. In Miami, for instance, as in other cities around the world, sea walls and rising highways have been the first choice. But some form of managed retreat is already happening in New Orleans, as in Hawaii, New York, and other coastal areas. As we've seen, after Hurricane Sandy in New York (October 2012), the State government made it possible for residents of Staten Island to sell their homes to the government at a fair market price. Speaking about climate change, Tom Larsen, Senior Director for Insurance Solutions with CoreLogic, said, "Managed retreat is one of the things we have in our toolkit as we try to navigate the future. There will be some areas that are simply not rebuildable."[76]

In Houston, for example, one community with generations-long ties has been dismantled because of mandatory buyouts. A research team from Rice University found that government-funded buyout programs for relocation can deepen social inequalities. Vulnerable people can be most at risk from this adaptation to climate change.[77]

Amy Chester is the managing director of Rebuild by Design, which works to help communities build greater climate resilience. She points to the difficult decisions communities will have to make with climate change in an aging society:[78]

> Communities are suffering and we all collectively have a responsibility to do something about it . . . We're gonna have to decide together: which are the neighborhoods that we want to fortify and where are the neighborhoods that we want to retreat? And if we retreat, where are our community members gonna go, and how do we ensure that when community members move upland that they're not further gentrifying communities?

Searching for safety is the challenge for aging in an era of what must be called the great climate migration. A major environmental organization published a much-awaited piece on the best places to live to avoid climate change. The answer?

> There is no easy answer on the best places to live to avoid climate change . . .," not everyone has the resources and means to relocate, nor should people be forced to leave their homelands to avoid the consequences of humanity's dependence on fossil fuels and environmental degradation.[79]

The USA has historically lived in a "niche" where, despite geographic differences, all 50 states have been habitable. But climate change will give a different picture of where that niche is: "In the case of extreme warming . . . the niche moves sharply toward Canada, leaving much of the lower half of the USA too hot or dry for the type of climate humans historically have lived in."[80]

What are the factors that will drive this great climate migration? Factors include extreme heat and humidity; big wildfires; sea level rise; crop yields; and economic damage from climate change. Above all, the greatest risk is the compounding damage from all these threats together. How do we picture what our future could be? ProPublica, drawing on research from the National Academy of Sciences, has put together a composite set of maps and visual displays that show how these trends are connected.[81]

Managed retreat is a close relative of forced migration, and it's much bigger than the USA. It is coming during exactly the time when population aging is reshaping what America looks like. More than a fifth of the earth's population could be living with extreme temperatures by 2100, creating new pressure for forced migration from climate change.[82] The climate crisis could push between 3 and 6 billion people into migration of some kind by 2070—but the trend begins much sooner than that. In fact, it has already begun.[83] It is likely that nearly 1.4 million people will have immigrated to America in 2023, a third more than before COVID-19. Similar waves were expected for Canada and Germany, among other countries of the Global North. Some substantial portion of these immigrants are driven by extreme weather events and thus by climate change.[84] In short, "Think globally, but act locally." The story of forced migration and managed retreat is both an American story and a global story. We need to think about both.

5.2   David Brower: Eco-Elder

Benjamin Franklin said: "Work as if you were to live a hundred years, and Pray as if you were to die tomorrow." What makes this seemingly contradictory position possible for human beings in the last stage of life? How was it reflected in Franklin's own career as an elder presiding over the making of the U.S. Constitution?

We can find a clue to Franklin's call in the phrase "late freedom" (*Spaete Freiheit*) or the unique capacity for liberation in later life, as described by the Austrian gerontologist Leopold Rosenmayr (1983). Late freedom means to go beyond the kind of care or concern that is self-protective. It means moving toward a level of liberation that gerontologist Tornstam called gerotranscendence.

David Brower was the long-time Executive Director of the Sierra Club, America's most influential environmental organization. Later, in his old age, he became the founder of another important advocacy group, Friends of the Earth. David Brower lived according to the ethics of the Late Freedom which he expressed so eloquently. He died in 2000 at the age of 88, still laboring on behalf of future generations. Brower loved to quote the African proverb "We don't inherit the earth from our ancestors, we borrow it from our children."

Notes

1 The one indispensable book about water is Peter Gleick, *The Three Ages of Water: Prehistoric Past, Imperiled Present, and a Hope for the Future* (Public Affairs, 2023).

2 Zoya Teirstein, "The Most Significant Drought in 1,200 Years" *Grist* (Aug. 13, 2022).

3 Daniel Newman, "Less Water, More People Present a Growing Economic Challenge for the Driest Counties in the United States" *Economic Innovation Group* (Aug. 16, 2022) at: https://eig.org/drought-conditions-snapshot/

4 Teirstein, "The Most Significant Drought in 1,200 Years."

5 SciTech Daily, "Hidden Danger: Researchers Warn That Climate Change Can Put the Planet's Largest Reserves of Drinking Water at Risk" (Feb. 3, 2024) at: https://newsreadeck.com/article/scitechdaily/hidden-danger-researchers-warn-that-climate-change-can-put-the-planet-s-largest-reserves-of-drinking-water-at-risk/6adde2e7d432adcfd2170084aab70d5e/ See also D. Carr, et al., "Population Aging and Heat Exposure in the 21st Century: Which World Regions Are at Greatest Risk?" *Journals of Gerontology* (Apr., 2024) https://doi.org/10.1093/gerona/glae053

6 Yale Environment, "Drought Linked with Health Risks in Older Adults, Yale-Led Study Shows" at: https://environment.yale.edu/news/article/study-links-drought-with-human-health-risks

7 Natalia Dmietrieva, et al., "Middle-Age High Normal Serum Sodium as a Risk Factor for Accelerated Biological Aging, Chronic Diseases, and Premature Mortality" BioMedicine, *The Lancet* (Jan., 2023) at: www.thelancet.com/journals/ebiom/article/PIIS2352-3964(22)00586-2/fulltext

8 Salvadora Coral, et al., "Effects of Droughts on Health: Diagnosis, Repercussion, and Adaptation in Vulnerable Regions Under Climate Change. Challenges for Future Research" *Science of the Total Environment* Feb. 10, 2020;703.

9 Babak Fard, et al., "Evaluating Changes in Health Risk from Drought Over the Contiguous United States" *International Journal of Environmental Research and Public Health* Apr. 12, 2022;19(8).

10 Kate Yoder, "Climate Change Has Toppled Some Civilizations But Not Others. Why?" *Grist* (Oct. 16, 2023) at: https://grist.org/culture/climate-change-societal-collapse-explained/

11 Jeff Masters, "Ten Civilizations or Nations That Collapsed from Drought" *Wunderblog Archive* (2016) at: www.wunderground.com/blog/JeffMasters/ten-civilizations-or-nations-that-collapsed-from-drought.html

12 Diana Leonard, "California Is Drought-Free for First Time in Years. What It Means" *Washington Post* (Nov. 8, 2023) at: www.washingtonpost.com/weather/2023/11/08/california-is-drought-free-first-time-years-what-it-means/

13 Michael Marshall, "Did a Mega Drought Topple Empires 4,200 Years Ago?" *Nature* (Jan. 26, 2022) at: www.nature.com/articles/d41586-022-00157-9

14 Grist, "Parched" at: https://grist.org/series/drought-parched/

15 Kevin Dennehy, "Drought Linked with Health Risks in Older Adults" *Yale School of the Environment* (2017) at: https://environment.yale.edu/news/article/study-links-drought-with-human-health-risks

16 Somini Sengupta, "City Living, with Less Water" *New York Times*, "Climate Forward" (Apr. 29, 2022) at: www.nytimes.com/2022/04/29/climate/drought-water-scarcity.html

17 Joshua Partlow, "Arizona City Cuts Off a Neighborhood's Water Supply Amid Drought" *Washington Post* (Jan. 16, 2023) at: www.washingtonpost.com/climate-environment/2023/01/16/rio-verde-foothills-water-scottsdale-arizona/

18 Wyatt Myskow, "Arizona Announces Phoenix Area Can't Grow Further on Groundwater" *Inside Climate News* (June 1, 2023) at: https://insideclimatenews.org/news/01062023/arizona-phoenix-development-water/

19　Alex Hager, "The 'Power of Aridity' Is Bringing a Colorado River Dam to Its Knees" *Inside Climate News* (Dec. 14, 2022) at: https://insideclimatenews.org/news/14122022/colorado-river-glen-canyon-dam-drought/ See also "As the Colorado River Declines, Water Scarcity and the Hunt for New Sources Drive up Rates" *Inside Climate News* (June 17, 2023).

20　Henry Fountain, "How Bad Is the Western Drought? Worst in 12 Centuries, Study Finds" *New York Times* (Feb. 14, 2022) at: www.nytimes.com/2022/02/14/climate/western-drought-megadrought.html

21　NBC News, "2022 Was the Year of Drought" at: www.nbcnews.com/science/environment/2022-was-year-drought-rcna62410

22　Dana Nuccitelli, "California Just Had Its Worst Drought in Over 1200 Years, as Temperatures and Risks Rise" *The Guardian* (Dec. 8, 2014). See also "Causes of California Drought Linked to Climate Change" *Stanford School of Earth Sciences* (Sept. 28, 2014) at: https://pangea.stanford.edu/d7-archive/sesd7/news/causes-california-drought-linked-climate-change/

23　Grist, "We Broke Down What Climate Change Will Do, Region by Region" *Grist* (Nov. 29, 2022) at: https://grist.org/cities/we-broke-down-what-climate-change-will-do-region-by-region/

24　Fountain, "How Bad Is the Western Drought?"

25　David Knowles, "Finding Safe Haven in the Climate Change Future: The Southwest" *Yahoo Climate* (Nov. 26, 2022) at: www.yahoo.com/news/finding-safe-haven-in-the-climate-change-future-the-southwest-100032145.html

26　Elizabeth Kolbert, "Climate Change from A to Z. The Stories We Tell Ourselves about the Future" *The New Yorker* (Nov. 21, 2022) at: www.newyorker.com/magazine/2022/11/28/climate-change-from-a-to-z

27　Elena Shao, "The Colorado River Is Shrinking. See What's Using All the Water" *NY Times* (May 22, 2023) at: www.nytimes.com/interactive/2023/05/22/climate/colorado-river-water.html

28　Brian Richter, "Water Scarcity and Fish Imperilment Driven by Beef Production" *Nature Sustainability* (Mar. 2, 2020) at: www.nature.com/articles/s41893-020-0483-z.epdf

29　Joshua Parlow, "The Colorado River Drought Crisis: How Did This Happen? Can It Be Fixed?" *Washington Post* (Feb. 5, 2023).

30　"The Very Bad Math Behind the Colorado River Crisis" at: https://grist.org/drought/colorado-river-water-rights-california-arizona-fight/ See also "At Last, States Reach Colorado River Deal: Pay Farmers Not to Farm" at: https://grist.org/drought/colorado-river-deal-arizona-nevada-california-conservation-agriculture/

31　Fourth National Climate Assessment (2022) at: www.globalchange.gov/nca4

32　Bryan Walsh, "Americans Keep Moving to Where the Water Isn't" *Vox* (Aug. 28, 2022) at: www.vox.com/2022/8/28/23322006/climate-change-heat-wave-phoenix-drought-housing-population

33　Deborah Carr, et al., "The Sunbelt Was the Retirement Destination of Choice. That Was Before Climate Change" *CNN* (Sept. 1, 2023) at: www.cnn.com/2023/09/01/opinions/seniors-retirees-arizona-florida-climate-change-heat-wave-carr-wing-falchetta/

34　Jake Bittle, "Tomorrow's Water Is Being Hoarded Today" *Grist* (May 5, 2023).

35　Greg Dalton, "California's Water Paradox" *Climate One* (Jan. 13, 2023).

36　National Park Service, "Death Valley Experiences 1,000 Year Rain Event" (Aug. 7, 2022) at: www.nps.gov/deva/learn/news/death-valley-experiences-1-000-year-rain-event.htm

37　"Large Expanses of Cropland May Fail Altogether" at: www.thecooldown.com/outdoors/100th-meridian-west-east-climate-divide-shift/

38　Michael V. McGinnis, *Bioregionalism* (Routledge, 1998).

39 Erica Gies, "Smarter Ways with Water" *Nature* (Nov. 16, 2022) at: www.nature.com/articles/d41586-022-03648-x?utm_source=newsletter&utm_medium=email&utm_campaign=newsletter_axiosfinishline&stream=top

40 Statement to station KQED, reported in *Axios*, 00 years of that divine Hetch Hetchy water" (May 11, 2023) at: www.axios.com/newsletters/axios-san-francisco-5601367b-017b-4dee-8215-ff3edafda5c0.html

41 David Wallace-Wells, "There's Nowhere to Escape the Smoke from Wildfires" *NY Times* (June 7, 2023) See also: "There's No Escape from Wildfire Smoke" at: www.nytimes.com/2023/05/17/opinion/wildfires-smoke-pollution-distance.html

42 Deborah Lynn Blumberg, "My Retirement Heaven Is Turning into Hell" *NextAvenue* (Oct. 2, 2023) at: www.nextavenue.org/my-retirement-heaven-is-turning-into-hell/

43 Susan Garland, "Do You Really Want to Rebuild at 80?' Rethinking Where to Retire" *N.Y. Times* (Nov. 22, 2022) at: www.nytimes.com/2022/11/18/business/where-to-retire-climate-change.html?smid=nytcore-ios-share&referringSource=articleShare

44 John Wasik, "Where Can You Breathe Easy in Retirement? How to Find a Community with Clean Air" *Morningstar* (Sept. 23, 2023) at: www.morningstar.com/news/marketwatch/20230923302/where-can-you-breathe-easy-in-retirement-how-to-find-a-community-with-clean-air

45 Garland, "Do You Really Want to Rebuild at 80?"

46 https://aginginplace.com/

47 Briana Sacks, "California Plans Big Insurance Shifts as Climate Change Hits Home" *Washington Post* (Sept. 22, 2023) at: www.washingtonpost.com/climate-environment/2023/09/21/california-insurance-risks-fire-climate/

48 Kathleen Sullivan, "A Child's Christmas in Wales: Revisited" *Code Red and Me: Rethinking Everything* (Dec. 25, 2022).

49 For more on Pete Seeger, see Allan M. Winkler, *To Everything There Is a Season: Pete Seeger and the Power of Song* (Oxford University Press, 2009).

50 According to data from CoreLogic, "2021 Climate Change Catastrophe Report" at: www.corelogic.com/intelligence/2021-climate-change-catastrophe-report/

51 Daniella Arigoni, *Climate Resilience for an Aging Nation* (Island Press, 2023).

52 There are some insurance options for heat, but they're largely relevant for outside workers, not those planning for retirement. See: Jake Battle, "Extreme Heat Is Here. Can Insurance Help Protect Us?" *Grist* (Aug. 4, 2023) at: https://grist.org/extreme-heat/extreme-heat-is-here-can-insurance-help-protect-us/

53 Atiya Mahmood, et al., "Better Emergency Preparedness Can Protect Older Adults from Climate Change" *The Conversation* (July 19, 2022) at: https://theconversation.com/better-emergency-preparedness-can-protect-older-adults-from-climate-change-186511

54 Jenny Schuetz, "How to Protect Your Home—And Your Wallet—Against Natural Disasters" *Brookings* (Apr. 6, 2023) at: www.brookings.edu/blog/the-avenue/2023/04/06/how-to-protect-your-home-and-your-wallet-against-natural-disasters

55 See https://ejscreen.epa.gov/mapper/ and www.epa.gov/arc-x/your-climate-adaptation-search

56 "As Climate Fears Mount, Some Are Relocating Within the US" *Wired* at: www.wired.com/story/as-climate-fears-mount-some-are-relocating-within-the-us/

57 David Knowles, "Finding Safe Haven in the Climate Change Future: The Northeast" *Yahoo News: Climate Change Future* (Oct. 29, 2022) at: www.yahoo.com/news/finding-safe-haven-in-the-climate-change-future-the-northeast-090026421.html

58 Jon Hurdle, "As Climate Fears Mount, Some in U.S. Are Deciding to Relo-cate" *YaleEnvironment360* (Mar. 24, 2023) at: https://e360.yale.edu/features/as-climate-fears-mount-some-in-u.s.-are-deciding-to-relocate

59 Eliza Relman, "Middle Class Americans Are Moving Straight into Fire and Drought Because They Can't Afford to Live in the Cities That Are Safer from Climate Change" *Business Insider* (Aug. 15, 2023).

60 Julia Gilland Jenny Schuetz, "How to Nudge Americans to Reduce Their Housing Exposure to Climate Risks" *Brookings* (July 27, 2023) at: www.brookings.edu/articles/how-to-nudge-americans-to-reduce-their-housing-exposure-to-climate-risks/

61 "Is Compulsory Managed Retreat Our Future?" at: www.newamerica.org/future-land-housing/briefs/is-compulsory-managed-retreat-our-future/

62 "Coverage of Migration from Place to Place Within the U.S." *YaleEnviron-ment 360* at: https://yaleclimateconnections.org/2022/06/coverage-of-migration-from-place-to-place-within-the-u-s/

63 Samantha Allen, "30% of Americans Cite Climate Change as a Motivator to Move in 2023" *Forbes* (Oct. 9, 2023) at: www.forbes.com/home-improvement/features/americans-moving-climate-change/

64 "'It's Happening Now': How Rising Sea Levels Are Causing a US Migration Crisis" at: www.theguardian.com/environment/2022/apr/07/its-happening-now-how-rising-sea-levels-are-causing-a-us-migration-crisis Public response to climate migration is not simple: See Kate Yoder, "What Happens When You Read an Article About Climate Migration?" *Grist* (June 16, 2023) at: https://grist.org/language/climate-migration-study-articles-xenophobia/

65 Scott Wilson, "California's Cliffs Are Crumbling as Climate Change Reshapes the Coast" *Washington Post* (May 26, 2023) at: www.washingtonpost.com/nation/2023/05/26/california-coastline-changes-cliffs-climate-change

66 Craig Miller, "Choosing a Place to Retire? Factor in Climate Change" *NextAvenue* (Aug. 31, 2020) at: www.nextavenue.org/choosing-a-place-to-retire-climate-change/

67 Kent McClanahan and Angie O'Leary, "How Climate Change Could Impact Your Retirement Plans" *MarketWatch* (Oct. 5, 2021) at: www.marketwatch.com/story/how-climate-change-could-impact-your-retirement-plans-11633110207 See also Rodney Brooks, "The Impact of Climate Change on Retirement Planning" *USNews Money* (June 21, 2022) at: https://money.usnews.com/money/retirement/aging/articles/the-impact-of-climate-change-on-retirement-planning#:~:text=Climate%20Change%20and%20Your%20Retirement%20Budget,buy%20insurance%20for%20natural%20disasters.

68 Jake Bittle, *The Great Displacement: Climate Change and the Next American Migration* (Simon & Schuster, 2023). See also: Jake Bittle, "Florida's Great Dis-placement Has Already Begun" *Business Insider* (Feb. 21, 2023).

69 Susan Garland, "Do You Really Want to Rebuild at 80?' Rethinking Where to Retire" *N.Y. Times* (Nov. 22, 2022) at: www.nytimes.com/2022/11/18/business/where-to-retire-climate-change.html?smid=nytcore-ios-share&referringSource=articleShare

70 See "Climate Ready Communities" at: https://climatereadycommunities.org/learn-more/about-guidebook/

71 Jack Healy, et al., "When the Heat Can't Be Beat" *New York Times* (July 28, 2022) at: www.nytimes.com/2022/07/28/us/heat-records-summer-climate.html?referringSource=articleShare

72 Rachel Alexander, "How Do You Love What You May Soon Lose?" *New Ameri-can Weekly* (Feb. 14, 2019) and later in *B the Change* at: https://bthechange.com/how-do-you-love-what-you-may-soon-lose-80ef7e0ccdbf

73 David Knowles, "Finding Safe Haven in the Climate Change Future: The North-east" *Yahoo News: Climate Change Future* (Oct. 29, 2022).

74 George Manbiot, "The Cruel Fantasies of Well-Fed People" (Oct. 4, 2023) at: www.monbiot.com/2023/10/04/the-cruel-fantasies-of-well-fed-people/?utm_source=substack&utm_medium=email

75 Delger Erdenesanaa, "What Is 'Managed Retreat'?" *Inside Climate News* (2023) at: https://insideclimatenews.org/climate-101/what-is-managed-retreat/

76 "Weather Disasters Affected 1 in 10 Homes in the Country Last Year, Report Finds" at: www.washingtonpost.com/politics/2022/02/17/weather-disasters-affected-1-10-homes-country-last-year-report-finds/

77 "Torn Apart" *Grist* (Sept. 19, 2022), part of a series, "Flood. Retreat. Repeat" at: https://grist.org/series/flood-retreat-repeat/

78 Climate One, "Managed Retreat: When Climate Hits Home" *Climate One*, Podcast (Dec.23, 2021) at: www.climateone.org/audio/managed-retreat-when-climate-hits-home

79 Tin Lok Wu, "Best Places to Live to Avoid Climate Change" *Earth.Org* (Mar. 5, 2024) at: https://earth.org/best-places-to-live-to-avoid-climate-change/

80 "Climate Change Will Force a New American Migration" "The Great Climate Migration" *ProPublica* (2021) at: www.propublica.org/article/climate-change-will-force-a-new-american-migration

81 "New Climate Maps Show a Transformed United States" *ProPublica* (2021) at: https://projects.propublica.org/climate-migration/

82 Johnny Wood, "Climate Extremes Could Force Millions to Migrate" *World Economic Forum* (May 31, 2023) at: www.weforum.org/agenda/2023/05/climate-extremes-and-other-nature-and-climate-news-you-need-to-read-this-week/

83 "Climate Crisis Is on Track to Push One-Third of Humanity Out of Its Most Livable Environment" at: www.propublica.org/article/climate-crisis-niche-migration-environment-population

84 "A New Wave of Mass Migration Has Begun: What Does It Mean for Rich-World Economies?" *The Economist* (May 28, 2023). See also: "Climate Change Could See 200 Million Move by 2050" at: www.yahoo.com/news/report-climate-change-could-move-130115266.html

Part III
Ethics

6 Who Is to Blame for Global Warming? The Wealthy, Rich Countries, Big Business, Older People

> In a free society, some are guilty but all are responsible.
>
> —Rabbi A.J. Heschel

Start with a provocative question: "Who's to blame for global warming?"

"Blame." Blame means guilt. Is it true, as they say, that guilt is "the gift that keeps on giving?" Or is guilt just another variety of psychopathology, better to get rid of it as soon as possible? As a society, we're making great progress in eliminating shame. Why can't guilt be the next to go?

Guilt has a long pedigree: the Jews invented it; the Catholics perfected it; and the Protestants turned it into the work ethic. Rabbi Heschel wants us to understand that, even when some individuals are guilty, we are all responsible. Whether we call it guilt or not, this responsibility is absolutely shared when it comes to climate change.

I was brought up as a Protestant; spent much of my career in Jewish fundraising; and got an advanced degree in Catholic philosophy. I even give talks on Rumi and Islamic mysticism, so when it comes to religion I cover all the bases. I have enough guilt to go around and even to share with others.

But here's the problem. As a society, we're killing ourselves with the work ethic and that overactivity is often linked to guilt, ambition, and other unlovely traits. Overconsumption and waste are major drivers of the climate crisis. Work, consumption, energy use: it's all too much, and at some level in ourselves, we all know it.

Those who are guilty of damaging the climate, of our planet, should be guilty. Let's not let them off the hook, even if we find ourselves, too, among those "guilty, as charged." Let's not adopt the passive voice: for example, "Damage was caused . . ." Let's keep focused on who did it, when, where, why. Let's keep guilt, and blame, very much in the picture. Without guilt and blame, we won't have the action we need.[1]

The question about "Who's to blame for . . ." will offend some people. For example, Barbara Fried is the mother of Sam Bankman-Fried, convicted in 2023 of fraud, conspiracy, and money-laundering. Fried wrote in her piece "Beyond Blame" for the *Boston Review* that our society is engaged in

DOI: 10.4324/9781003345992-9

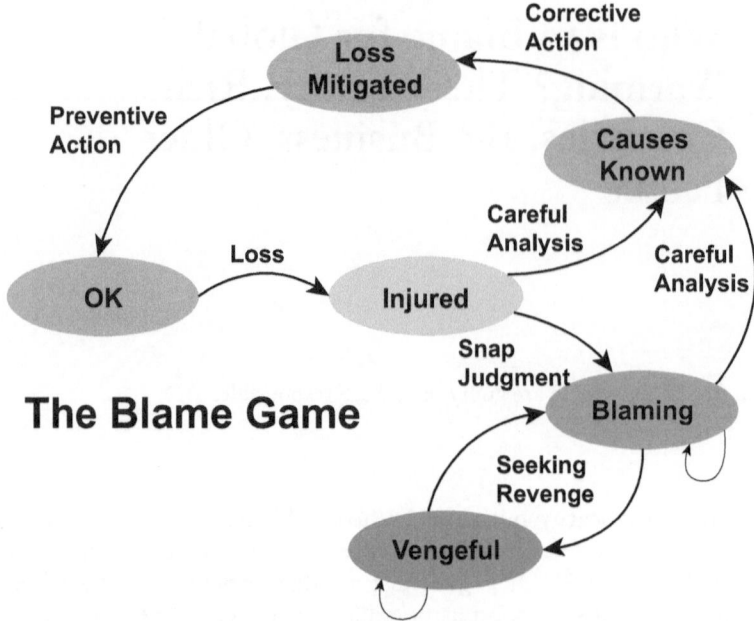

Figure 6.1 The Blame Game. Public domain.

a "blame fest," giving us nothing "except the guilty pleasure of reproaching others for acts that, but for the grace of God, or luck, or social or biological forces, we might well have committed ourselves."[2] We can certainly apply this talk about blame for climate change. Think about the multiple ways in which those with any modest degree of wealth are, day by day, to be blamed for burning carbon and putting it into the atmosphere. Are we all beyond blame for any of it?

Talking about blame in the ethics of climate change here is quite intentional. It's reasonable to ask questions like "Who was to blame for racial segregation?" or "Who is to blame for child trafficking." We will come up with different answers, and those blamed can be individuals or groups of people, which introduces the difficult question of collective responsibility and guilt. But I want us to pose the question this way, which is also why the term "global warming" is used here. It is not evasive like the term "climate change" itself, which blurs responsibility, as perhaps it was intended to do by the PR consultants who originally came up with the term. Those in denial say "The climate is always changing, isn't it?" My whole point about climate change is to say: No excuses.[3]

The question of "Who's to blame for global warming?" reminds us that the climate catastrophe is not a painful circumstance in which we find ourselves without any clear cause. Climate change is not like earthquakes. We don't ask who is responsible for earthquakes because earthquakes aren't caused by human agency. Climate change has a human cause, so questions of responsibility are paramount. Anyone not deluded by climate denial now understands the human role. For just this reason we need to ask troubling questions, including the question of blame. We could enlarge it to include the question of "Who's to blame for climate denial?" but leave that aside for the moment. The point of this dialogue is to understand our responsibility, not to engage in a blame game.[4]

We naturally think of fossil fuel companies, rich countries, rich people, political leaders, and maybe all of us who elected them. And then we have answers to our question, Who's to blame for global warming?[5] When we speak about climate change and aging, it's also natural to blame older people. Doesn't it make sense that I, at age 78, am more responsible for the climate crisis than my granddaughter at age 5? We will return to this question of generational justice soon enough.[6] So, yes, elders like me are also responsible. We are not blameless.

We who are old need to face our own responsibility right up front, in the words of Bill McKibben, founder of Third Act:

> Young people are currently at the forefront (on climate change), and that makes sense: they're going to have to live out their lives on a heating planet. But it's not practical to assign the deepest dangers we've ever faced as a species to 17-year-olds as if it was extra homework. And it's not fair either, because they didn't cause it. We did— those of us of a certain age . . . If you're 70 years old, 85% of the planet's carbon emissions have occurred in your lifetime. Yes, climate change had its birth in the 1700s when we first learned to burn coal in engines, but the process really got going in our lifetimes, and in the United States . . . By 1970, Americans consumed a third of the world's energy—more than the Soviet Union, Britain, West Germany, and Japan combined.[7]

So, who is responsible? Here's the answer you'll find in this chapter. We all are to blame, we all are responsible but not equally so. Rabbi Heschel put it best when he said, "In a free society, only some are guilty, but all are responsible." But, again, we are not equally responsible. In the pages that follow we will look closely at candidates in our criminal "line-up." In looking at that line-up we'll see human agents who need to be held guilty, both individuals and groups. In the last scene of the film *Casablanca*, Capt. Renault said, "Round up the usual suspects." We're familiar with these suspects, whether or not we use the language of blame. Those usual suspects are fossil fuel companies, rich people, rich countries, and others who have massively contributed to the warming of the planet in the past 30 years. Why the past 30 years? Because there's another group who also belongs in this line-up. They are in the title of this book: those who are old.[8] To see one of those responsible, at age 78, I need only to look in the mirror.

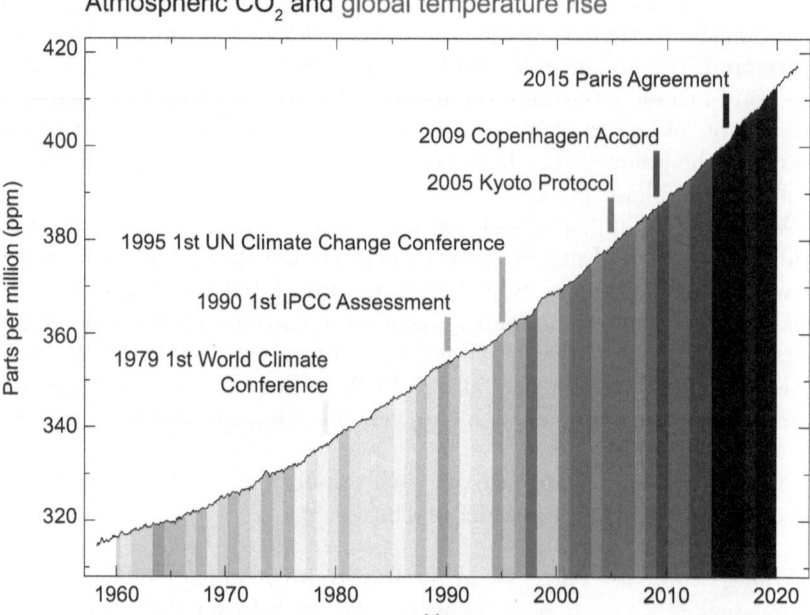

Figure 6.2 Trends in atmospheric CO_2 and global temperature change and climate policies. Public domain.

How can I say I didn't know about the threat of climate change?[9] As a teenager, I was a Boy Scout, working at camps Wauwepex and Onteora in 1960, just before I got my Eagle Scout Award. The influential *The Limits of Growth*, by Donella Meadows and colleagues at MIT, was published in 1972. That book made an impact on me, so much that in 1975 I taught a course on "The Crisis of Civilization" at New York University and was able to meet and interview environmentalist Lewis Mumford. A decade later I was finally able to meet personally with Alexander King, who had been head of the Club of Rome when his group sponsored *The Limits of Growth*. Looking back four decades—to 1983—shows an early government with greenhouse warming. The Environmental Protection Administration (EPA) warned about climate change. It also shows just how long the world has been dragging its heels.[10]

Wasn't I reading the newspapers in 1988 when James Hansen gave his pivotal testimony on global warming to Congress? Did any of these things put a dent in my habits of driving in cars, of flying in airplanes, of living to the peak of the fossil fuel world in which I grew up and flourished, like all the Boomers around me? How can I say, "I didn't' know?" I was 30 years old when I started teaching about climate change. But it took 45 years until I became concerned enough to pick up the thread again and begin working on this book. I have to consider myself a specialist in denial.

James Hansen later became director of the climate program at Columbia University's Earth Institute. Hansen has said the best hope was for a generational shift of leadership. He now wants young people to take up leadership to lead us away from climate disaster: "The bright side of this clear dichotomy is that young people may realize that they must take charge of their future. The turbulent status of today's politics may provide opportunity."[11]

Another confirmation comes from Richard Heinberg (born 1950), who is an aging Boomer and Senior Fellow with the Post Carbon Institute. He's a long-time environmental advocate, and I've had a chance to get to know him personally. In 2022, he offered this message:

> My generation basically had all the information it needed. We had books like Silent Spring, Limits to Growth, and Small Is Beautiful. We could have tamped down consumption, but instead, we threw the biggest party in all of human history and we're leaving the next generations to clean up after us.[12]

LET THEM NOT SAY

by Jane Hirshfield

Let them not say: we did not see it.
We saw.
Let them not say: we did not hear it.
We heard . . .
Let them not say: they did nothing.
We did not-enough.
Let them say, as they must say something:
A kerosene beauty. It burned.
Let them say we warmed ourselves by it,
read by its light, praised, and it burned.[13]

Jane Hirshfield, in "Let Them Not Say," is reminding us, like Rabbi Heschel, that we cannot escape responsibility: the example that she gives, "kerosene beauty," invokes fossil fuel to make the point. Her message is addressed to future generations.

Do we actually care about future generations? That's what action around climate change demands of us elders. Can we bluntly admit that we're failing that challenge? Canadian David Suzuki, age 87, put it best: "We've failed big time." In an early scene in Jean-Luc Goddard's 1960 film *Breathless*, one character challenges the hero, Jean-Paul Belmondo, by saying "Mister! You don't have anything against youth, do you?" In defiant style Belmondo replies "Yes, I do. I prefer old people." As we burn more and more fossil fuels each year, the burden on future generations grows ever higher: "Driven largely by soaring use of cheap fossil fuels, our energy consumption has increased sixfold since 1950. Since that date, we've consumed about 60% of all the energy we've produced in our species' existence."[14] Old people, like me, have enjoyed our life. We won't be here much longer, but we still have obligations. Can we measure up to the challenge? Speaking from an Indigenous perspective, Robin Wall Kimmerer speaks directly to that challenge[15]

> The moral covenant of reciprocity calls us to honor our responsibilities for all we have been given, for all that we have taken. It's our turn now. Whatever our gift, we are called to give it and to dance for the renewal of the world.

Can't we recognize that our whole society was warned? The warnings go back a long way.[16] Eunice Newton Foote, a scientist and women's rights advocate in Seneca Falls, New York, warned about it as early as 1856. In 1896, there was an explicit warning that burning fossil fuels could promote global warming of the earth. That warning came from Svante Arrhenius, a Nobel-prize-winning Swedish chemist, and a distant relative of Greta Thunberg. In the 1950s oceanographer Roger Revelle, at Harvard, picked up the warning and was able to educate one of his undergraduate students, Al Gore, who would go on to make the film *An Inconvenient Truth* (2006).

There are those who are against blaming anyone for climate change. Glen Peters, who directs research for the Center for International Climate Research in Norway, takes that view. Blaming people will not be fruitful. Is he right? Or isn't it essential that we talk about blame—or accountability or responsibility or whatever terms we want to use? The point is that climate change is a moral challenge, not just a technical problem. Whether we label it blame or not, the question of who is responsible for the climate crisis is a necessary one. It will inevitably impact the solutions we propose to fix things. But, as noted, pinning down who to blame is not an easy job.

Who says guilt is a bad thing? Lots of people say that today, don't they? How often do we hear that we should avoid being "judgmental," as in that saying "Guilt—the gift that keeps on giving." We can—and should—recognize that "We're all at fault" when it comes to climate change. But we're not all *equally* at fault. That's where guilt—appropriate guilt, let's call it—comes into play. For example, it turns out that oil heirs are funding disruptive climate protests.[17] And appropriate guilt is the reason, to which I say: great—let's have more guilt, guilt, and blame in the right places.

But what about "carbon guilt" shared more widely? Aren't we all guilty? At least those of us who drive in cars, live in concrete buildings, and eat a typical diet. Yes, that's why I said "all of us." Consider *Climate Change as Class War* by Matthew T. Huber and its discussion of "carbon guilt" and consumption. Huber makes the argument that this cycle of guilt is only attractive to the professional classes but completely fails to address that it is actually the producing class who is causing climate change. Huber points the finger at producers, not consumers, including the fossil fuel industry, aviation industry, and steel industry. According to a major study, now a decade old, nearly two-thirds of the greenhouse gas emissions come from fossil fuel use, methane leaks, and cement manufacturing. These greenhouse gas emissions came from just 90 companies across the world.[18] It sounds like we're just rounding up the usual suspects!

Carbon guilt can be targeted very broadly: in fact, all around the world.[19] Closely related is the concept of "carbon footprint," leaving us with the question of how much should I feel guilty about my carbon footprint: A carbon footprint is the amount of carbon dioxide released into the atmosphere through the actions of an individual, organization or country. The idea of a footprint came from two Canadian researchers in the 1990s as a metaphor for humanity's impact on the plane.[20]

The idea of carbon footprint by individuals naturally leads to commandments for individual virtue: Turn off lights. Eat less meat. Walk to work. Fly less. Buy less. Recycle. It turns out that the idea of carbon footprint by individuals is only the latest effort by producers to shift attention from producers to consumers. It's a well-worn path, pioneered by plastic producers ("Make America beautiful") and later by tobacco companies and now by fossil fuel companies.

In short, we can't avoid the central challenge:

Voluntary carbon offsets allow people to invest in projects that allegedly counteract their greenhouse gas emissions. But can voluntary offsets help slow global warming? Or are offsets simply a way for guilt-ridden consumers to buy their way out of bad feelings?[21]

I have a colleague, an ethicist who is also a leading member of Third Act, the climate advocacy group whose members are people over the age of 60. When we talked about why there should be a group of climate advocates coming together based on age, he said, quite cogently, that it's because they have money and because they vote. He was right, of course. To those who have been given much, from them much shall be asked. Here is a major argument of this book, as well as for the existence of groups like the Third Act and Elders Climate Action. There are many older people who already recognize their responsibility and their power to make a difference.[22]

The point is NOT to blame elders as prime actors in the climate catastrophe. When we do that, it is all too easy to fall into the habit of ageism, a term invented by my mentor, Dr. Robert Butler in 1969.[23] Blaming the old as the big agents for climate change too quickly becomes a way of pitting the old against the young when youth and elders need to be allies in the climate emergency we face.[24] If we ask whether elders are "victims or villains'" maybe we have just gotten the question wrong.[25] We live at a moment—the next few decades—when climate change coincides with population aging, so, yes, elders are responsible:

> In addition to the fact that there will be more people over the age of 65, a general increase in the numbers of warm days and nights as well as the frequency of extreme hot temperatures expected in the U.S. [. . .] Together, the ageing of U.S. population and the changing climate may also intensify the energy demand among the elderly.[26]

So, yes, age matters.

Elders are not blameless and they are not helpless, as in the concluding lines of Alfred Lord Tennyson's *Ulysses*, depicting Ulysses in old age:

> Tho' much is taken, much abides; and tho'
> We are not now that strength which in old days
> Moved earth and heaven, that which we are, we are;
> One equal temper of heroic hearts,
> Made weak by time and fate, but strong in will
> To strive, to seek, to find, and not to yield.[27]

There is a temptation to respond to climate change by getting young people angry with their parents and grandparents, if young people believe that elders are not bothered by the staggering upward distribution of money taking place around us.[28] Convincing the young to blame the old, instead of the rich, is a mistake. But it is also wrong to imagine that young and old are equally responsible for the way the world is, including the climate crisis. The young are not mistaken that they will be living on a planet left to them by elders, both rich and poor, who enjoyed the benefits of fossil fuels. The mistake is to create a false dichotomy here: if the old are responsible, then we let off the rich. If the rich are responsible, then we let off the old. We need to get over this and other false dichotomies: USA versus the world, old versus young, rich versus poor.

Nonetheless, it remains true that a 78-year-old, like me, is more responsible than my granddaughters, aged 4 and 6. I have simply lived longer, I've taken more global airline flights, I have more money, and I'm a bigger part of the consumer economy. People in Europe and America, the Global North, have contributed far more to global warming than people in Bangladesh, the Pacific Island countries, or others in the Global South. Yet those who have contributed the most to the climate burden will bear the least. Those who have contributed the least will be punished the most. Here, in stark terms, is the ethical problem at the center of this question, "Who's to blame for global warming?" The ethical condition of our time is exactly upside down. Those who have done the least will be punished the most. That's why there needs to be actions by elders who work for the world they are leaving behind. Along with wealth and power, along with individual and collective responsibility, the matter of age and the life course must be part of the conversation about who is to blame for global warming.

As we look at the "usual suspects" responsible for our situation, we will conclude with age and face the challenge of overcoming objections by elders around action on global warming. The discussion will be the basis for successive chapters in this book where we look at specific areas where people of any age, including those who now are old, can act in response to climate change. These include actions like becoming a climate-conscious consumer, investing for a green retirement, and working as citizens for public policy in a world of climate change. But let's begin with the usual suspects.

Are Corporations to Blame?

When we think about companies responsible for greenhouse gas emissions, we naturally point the finger at fossil fuel companies.[29] We need to ask the questions that senators asked when President Richard Nixon was being investigated for impeachment: about the companies, What did they know, and when did they know it?

The questions are important. If fossil fuel companies are to blame, then maybe they should be paying for the damage caused. An important study looked at the damage done by the oil, gas, and coal industries and came up with a "conservative" $23 trillion per year in damage to the earth's climate. It's possible to get beyond collective guilt and to identify the 21 largest fossil fuel companies that would owe a total of $209 billion per year in reparations for climate change. Just 100 companies are responsible for more than 70% of global emissions[30] "Who is to blame for global warming?" is a question that may have some answers.[31]

California Attorney General Rob Bonta (D) said, "Oil and gas companies have privately known the truth for decades—that the burning of fossil fuels leads to climate change—but have fed us lies and mistruths to further their record-breaking profits at the expense of our environment." Bonta concluded, "Enough is enough." California, a top oil-producing state, has now been suing Big Oil, thus becoming the latest and largest player among a growing number of governments looking to hold fossil fuel companies accountable for the effects of climate change. Former Vice President Al Gore pointed to "what the fossil fuel industry doesn't want you to know."[32]

Just 20 companies were responsible for one-third of global greenhouse gas emissions from fossil fuels and cement in 2018. These include Saudi Aramco, Chevron, Gazprom, ExxonMobil, National Iranian Oil, BP, and Royal Dutch Shell.

Climate activists demand change from fossil fuel companies. But more are also demanding change from financial groups that provide them with capital.[33] But blaming fossil fuel producers themselves neglects the other side of the equation: consumers of fossil fuels. It also neglects other industries that are huge sources of carbon pollution. And it neglects the role of governments in providing subsidies to fossil fuel producers, subsidies that the International Monetary Fund has estimated amount to $11 million each minute.[34] Governments are still pouring $7 trillion into subsidies for fossil fuels, and the true cost of climate pollution amounts to 44% of corporate profits.[35]

For example, cement is used in buildings and infrastructure all over the world. Manufacturing cement accounts for at least 8% of global CO_2 emissions. The Carbon Almanac speaks about the "Four C's" of climate change: along with coal, combustion, and cows, there's also concrete.[36] "Flight shaming" has become a moral position for those concerned about climate change, but it's oversimplified. We easily blame airlines for their role, but aviation contributes to less than 3% of all global emissions, according to the International Energy Agency. Greenhouse gas emissions from cement are big, and it's getting worse. From 2002 until 2021, the cement industry's emissions more than doubled: from 1.4 billion tons to 2.9 billion tons, according to the Center for International Climate Research. Cement production is now the second-largest emitter of carbon dioxide.[37]

Environmental activists have recognized the problem. By now over 50 companies, including big players like Google and Salesforce, have committed themselves to buying "low-carbon" cement, along with aluminum and steel, among other construction elements with a high carbon footprint. Climate advocates can point the finger at fossil fuel companies. But to find other major agents to blame for climate change, we only need to look around us in the buildings in which we live.[38]

Fossil fuel companies and other major polluters aren't just the source of greenhouse gas emissions. They're also the prime agents for confusing the public and preventing action on climate change. John Cook, a leading researcher on climate change communication, puts it this way:

> Ultimately, its goal is to delay climate action and maintain the status quo. And you will find that no matter what the argument is, the conclusion is always the same: whether they're arguing climate change isn't real, therefore we shouldn't act. Or climate change isn't caused by humans, therefore, we shouldn't act. Or solutions won't work, therefore we shouldn't act . . . what's most potent right now is culture war type misinformation . . . Arguments that other people who care about climate change, who are trying to get climate action . . . painting them as different to us and they're trying to take away our lifestyle or impinge on our freedom . . . The more tribal and polarized it becomes the harder it is to get progress.[39]

What about other companies beyond fossil fuel producers? The Business Roundtable is the leading voice for corporate America, and it has said that it believes "corporations should lead by example [and] support sound public policies . . . needed to address climate change." But it opposed the landmark climate legislation of our time, the Inflation Reduction Act, because companies feared higher taxes. They didn't even mention climate change. In 2009, when Steve Jobs was CEO of Apple, the company resigned from the U.S. Chamber of Commerce because of the chamber's feeble position on climate change. However, the chamber opposed the Inflation Reduction Act and its strong steps against climate change.[40]

By 2023, there was wide public debate about artificial intelligence, such as ChatGPT and other varieties of digital communication. But fossil fuel companies and other major polluters have been in the game already:

> "We have to recognize that AI is already here. As annoying as it is to admit, AI in its various forms will be a fundamental part of the planetary landscape for all the futures we can imagine. Or rather, for all the futures being imagined for us by the people who keep making advanced tools that both lift and crush ordinary human lives."[41] Here's the irony: "The very corporations that designed and funded campaigns to deceive the public and block climate action for decades now want the public to trust them. And they expect us to believe not only that they're part of the solution to climate change, but that we can't accomplish the transition to clean renewable energy without them."[42] We need to face the reality that the configuration of the private marketplace is what produces the insecurity afflicting people—above all the insecurity that we call climate change[43]

Rich and Poor Countries

When we look around the world, we recognize that the Global North—North America, Europe, and Japan—are only 12% of the global population, but these are the places that have produced half of the greenhouse gases since the Industrial Revolution.[44] When it comes to greenhouse gas emissions, the USA produces 18 tons per person, while, in India, the amount is lower than 3 tons.[45]

"The Global South—low or middle income countries in Africa, Asia, Oceania, Latin America and the Caribbean—is suffering disproportionately." The contrast is stark: in 2020, all of sub-Saharan Africa had a carbon footprint of a tenth of 1 ton compared to more than 15 tons in the USA.[46]

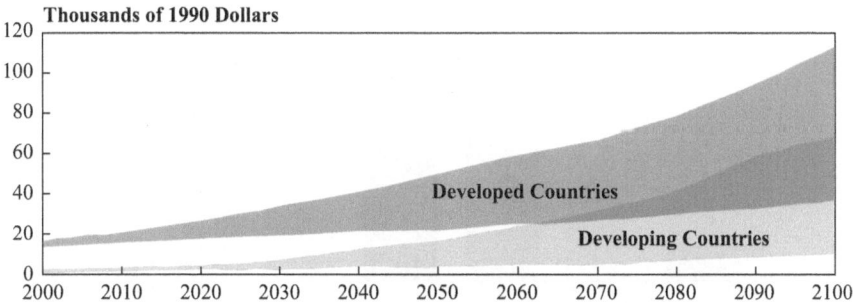

Figure 6.3 Range of uncertainty in gross domestic product per capita for developed and developing countries. Source: Congressional Budget Office based on Nebojša Nakićenović and Rob Swart, eds., *Emission Scenarios* (Cambridge, U.K.: Cambridge University Press, 2000). Public domain.

We can pinpoint the high-emitting countries by name, from satellite data. The biggest emitters, in sheer quantity, are China, the USA, India, Indonesia, Malaysia, Brazil, Mexico, Iran, Japan, and Germany.[47] Not all countries are equally responsible. In answer to the question of this chapter, wealthy countries are more to blame for global warming.[48]

What would it cost to balance this? $200 billion a year: that is, the amount of money underdeveloped countries would need each year to cope with the impact of climate change.[49] Is it a surprise, then, that countries around the world announce "net-zero" targets without actually coming up with reliable pledges or actions to reach the goal?[50]

Maybe rich countries should pay the poor for what's already been done. Needless to say, countries that are the big greenhouse emitters, both in the Global North and in the developing world, are not eager to pay reparations for the climate crisis now underway. The United Nations has warned that failing to provide that kind of money will mean suffering and, very likely, increased conflict. Climate justice means rich countries must respond to the needs of the developing world.[51]

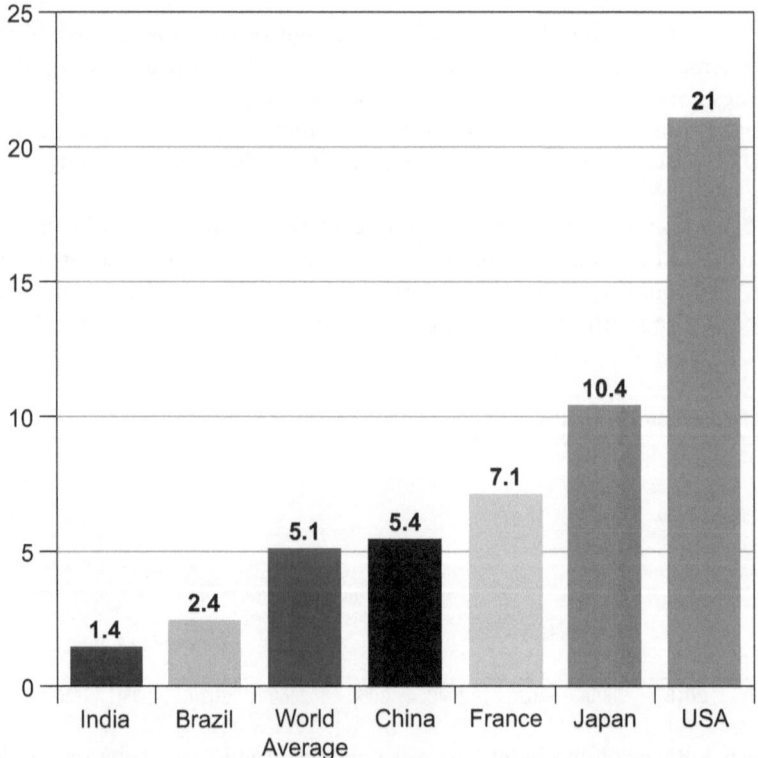

Figure 6.4 Global carbon dioxide emissions (tons per person annually). Adapted from a graphic by the Union of Concerned Scientists.

6.1   Alexander King: Eco-Elder

Figure 6.5 Alexander King. Public domain.

Alexander King was a distinguished British scientist well remembered as a key figure in the writing of the book *The Limits of Growth*, published in 1972. The book was translated into 30 languages and sold 30 million copies. One of those copies came to my attention in the year 1973, and it made a decisive impact on my thinking. It was not until many years later, in 1990, that I had the opportunity to meet Alexander King, because it turned out, earlier unknown to me, that he was a lifelong friend of Lawrence Morris, a friend of our family who ended up living in our house until he died at the age of 97. My job in 1990 was to drive Larry Morris to meet with Alex King, who continued his work as an Eco-Elder into advanced old age, when he died in 2007 at the age of 98.

Alex King did not write *The Limits of Growth* himself but, along with Italian industrialist Aurelio Peccei, King co-founded the Club of Rome in 1968 and later became President of this global think-tank. Even into advanced old age, he remained a major figure warning about the impact of unregulated economic growth. At first, King was a lonely voice in a time of optimism about the future of technology and industrial growth. But, like Larry Morris, who had also been a government official, Alex King had another side to his life.

He had over many years an abiding interest in spiritual growth and dimensions of human consciousness not grasped by the conventional mind. In short, King had a deeper contemplative life that coexisted with his active life and his public career.

The Limits of Growth was authored by Donella Meadows, with coauthors, and has had a vast impact on environmental thinking in the half-century since it first appeared. Meadows died too young, but even before the climate crisis was fully recognized she said there is too much bad news to justify complacency and there is too much good news to justify despair.

For me *The Limits of Growth* proved a catalyst for my own thinking about how human civilization, and even the human species itself, can flourish in a time of climate change.[52]

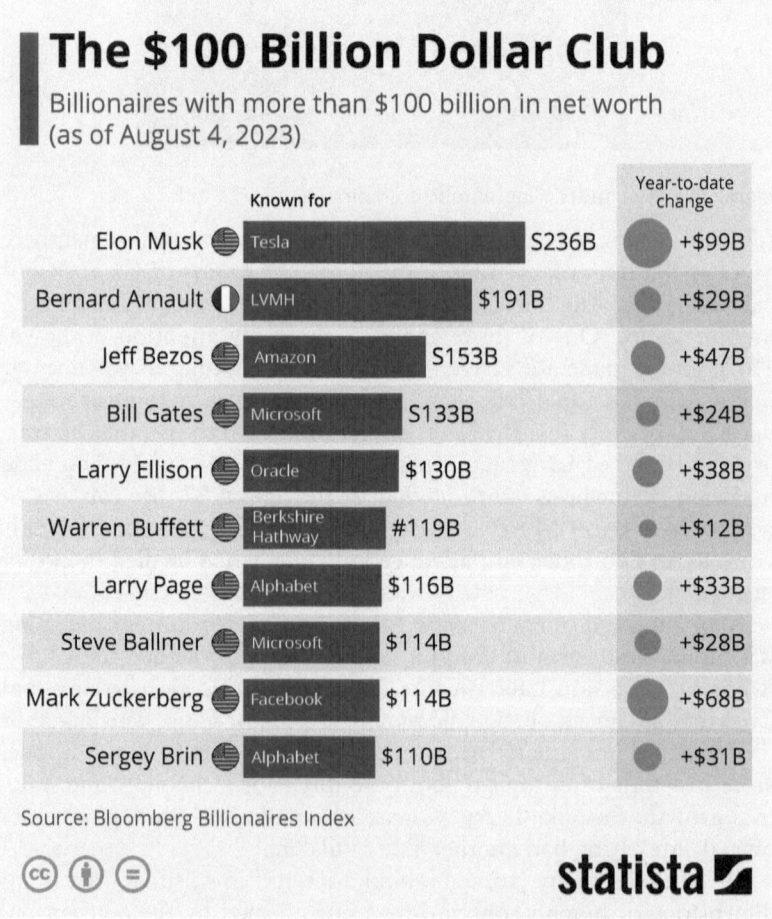

Figure 6.6 Increase in billionaire wealth (1987–2022) in U.S. dollars (trillion real terms). Adapted from a graphic by Statista.

Are Rich People Responsible for Climate Change?

Well, the figures seem to confirm the point. The top 1% of humanity can be blamed for twice as much greenhouse gas emission as the poorest 50%. By the end of this decade, the carbon footprint for the 1% will be 30 times what is required to limit global warming to 1.5% Centigrade.[53] Oxfam has estimated that the world's richest 1%, or 63 million people, produce double the greenhouse gas emissions of the world's poorest 3 billion people.[54]

When we look at America it turns out that 40% of U.S. climate emissions can be attributed to the richest households: that is, the top 10%. Within that group—which likely includes many, many readers of this book—there is also the top 1%. This wealthiest group was responsible for 16% of greenhouse gas emissions.[55] So, yes, the rich are to be blamed for climate change.[56]

Who are these rich people? Start with familiar names like the billionaires Bill Gates, Jeff Bezos, Mark Zuckerberg, Elon Musk, and Warren Buffett. We all know their names. And there are others, including women, less familiar, like Jacqueline Mars or Julia Koch. When we include their personal activities and their investments, it turns out that the world's richest people are responsible for a vast amount of greenhouse emissions.[57] To put it into numbers, 125 billionaires produce the same carbon footprint as the entire country of France, which has the seventh-largest gross domestic product of any country in the world.[58]

What are these rich people doing that is so bad? Let's start with private planes and yachts.[59] It's been said that operating a superyacht is perhaps the most damaging thing a single person can do to the earth's climate. The journal *Sustainability* calculated the impact of the yacht owned by DreamWorks co-founder David Geffen. It produces greenhouse emissions nearly 800 times what an average American produces in a year.

And things are getting worse. Economic research shows that the top 0.00001%'s share of total wealth today is ten times what it was just 40 years ago. The super-wealthy, with private airplanes and other carbon-producing assets, contribute disproportionately to greenhouse gas emissions. As the economic data shows, these trends are getting worse.[60]

The question about rich people and climate change is illustrated by the use of private jet planes. In a recent year, private aviation produced nearly as much as annual greenhouse emissions from Ireland. No wonder the Amsterdam airport decided to ban private jets as a step to protect our climate. We're getting closer to answering the question, Who's to blame for global warming? It turns out that a private jet will release 2 tons of carbon dioxide each hour. That's the equivalent of several months of greenhouse gas emissions, according to the European NGO Transport & Environment. Private planes are around ten times more polluting than commercial jets per passenger, and private planes are 50 times more polluting than high-speed rail.

What about the aviation sector as it whole? Air transportation is nearly 3% of global greenhouse gas emissions, but the proportion of the population responsible for it is very small: only 1% of the world's population produces half of carbon dioxide from commercial aviation, according to research published by the *Global Environmental Change Journal*. In short, air transport underscores the point: rich people are enormously responsible for global warming.[61]

For a long period after WWII, income and the gross domestic product showed a similar growth pattern. But, from around 1975, income for the lower 90% grew more slowly than the total economy, while income for the top 10% grew faster and continues to do so.[62] This is exactly the period when the super-rich were producing a disproportionate amount of greenhouse gases. So much for the top 1%.

Well, readers of this book may think, Oh, I'm not in that top 1%. But what about the top 10%? Those are individual Americans earning, on average, $248,000 or more (2022 dollars), and they produce 56 tons of CO_2, more than double what the richest 10% produce in Europe or China.[63] But global warming is exactly what it says: it's global. If we look at income and wealth from a global standpoint, then if you're earning more than about $100,000 each year, you're in the top 10% on a global scale.[64] How many readers of this book are close to that level?

We have looked at searching for safety when it comes to climate change. But searching for safety also applies to the very rich. They feel threatened by climate change, too. Prominent billionaires like Elon Musk and Jeff Bezos are fascinated with space travel, and they have their reasons. They often argue that human survival may depend on relocating to another planet in the event of environmental disasters, such as climate change.

As one who grew up on science fiction, I was at first intrigued by such interest in protecting the home planet by space travel. But further reflection makes it clear that, in this case, as in others, the very rich are perpetuating a climate nightmare.[65]

Whether or not space travel makes any sense, the richest of the rich are already concerned with protecting themselves. For instance, Douglas Rushkoff's book *Survival of the Richest*[66] gives a powerful critique, based on a story of five mysterious billionaires offering their views in a setting as a private desert resort. They're concerned about how to survive "The Event" or a global catastrophe that could happen if the climate crisis continues its disastrous trajectory. Like the rest of us, they're traveling on a fossil fuel merry-go-round and they don't want it to stop. If their programming technologists can't remake the world, then they will find ways to insulate themselves from disaster, protected as long as their own security guards don't turn against them. Capitalism has been around for a long time, but the degree of inequality today is beyond anything in prior times.[67] The tech billionaires depicted in Rushkoff's book are working on escape fantasies that leave the rest of us to the destruction unleashed by fossil fuel exploitation unchained. Searching for safety is—literally—out of this world.

The typical worker's wage has stagnated for the past 40 years while most of the economy's gains have gone to the top. Do we believe that people who are rich are succeeding because of their inherent worthiness or because the game is rigged in their favor? Have people who are poor failed, or has the system failed them? Is it morally acceptable that the pay of American CEOs has gone from an average of 20 times that of the typical worker 40 years ago to over 300 times today? These questions will be asked at a time when extreme weather events and climate change make more and more Americans dissatisfied with the world in which they are living.[68]

We have moved from looking at collective phenomena—countries and corporations—responsible for global warming and have looked at rich individuals, leading to the recognition of the importance of inequality as a structural dimension of climate change.[69] The fossil fuel system concentrates wealth in the hands of fewer people, and this economic framework always needs to be kept uppermost in mind. We now need to consider how climate change is reflected over the life course, from youth to old age.

Young People, Climate and Aging

It's appropriate that a book on climate change and aging should give attention to the youth. Young people can be the vital energy for climate advocacy, and they are demonstrating that already. The Oxford English Dictionary's Word of the Year in 2017 was "youthquake." It's what we need now, and, for it to work, it needs to be cross-generational: elders and youth together. Greta Thunberg was *Time Magazine*'s Person of the Year in 2019. But her campaign then was only beginning. By 2020 the Oxford Dictionary's Word of the Year was "climate emergency." It hasn't stopped being an emergency since then. Young people play a central role in social movements, over and over again: the U.S. Civil Rights Movement, the protest against the Vietnam War, the Women's Movement, Tiananmen Square in China, the Arab Spring, and Black Lives Matter. Official philanthropy has a long way to go in recognizing climate advocacy: climate change mitigation got less than 2% of global philanthropy in 2021.

We increasingly speak of a "climate emergency" recognizing that what is happening is not a gradual problem that we can ignore or something that will happen in the distant future. Climate is already an emotional burden for members of Generation Z born between 1981 and 1996.[70] That means my own son and daughter. Young people understand that the climate is a moral imperative: "You are stealing our future" is one version of "The personal is the political"— a slogan aging Boomers knew in the Sixties. For young people the stakes are greater if we fail to act. When children—and grandchildren—are calling for help, it's harder for elders to ignore the call. When speaking about the future, we should understand the real point, as Katie Elder put it: "Kids may be only 25% of the population, but they're 100% of the future." The Sunrise Movement, and its Green New Deal agenda, were influential in the 2020 elections.

The public opinion data are very striking when it comes to attitudes about climate among youth and older people. This data from the Pew Memorial Trust tells the story:[71]

Younger U.S. adults more open than older adults to phasing out fossil fuels completely

% of U.S. adults who say that the U.S. should ...

- ■ Phase out the use of oil, coal, and natural gas completely, relying instead on renewable sources
- ■ Use a mix of energy sources including oil, coal, and natural gas along with renewable sources

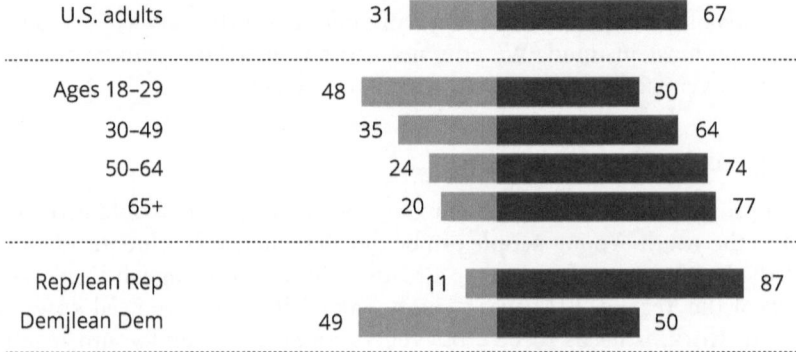

	Phase out	Mix
U.S. adults	31	67
Ages 18–29	48	50
30–49	35	64
50–64	24	74
65+	20	77
Rep/lean Rep	11	87
Dem|lean Dem	49	50

Among Rep/lean Rep who are ...

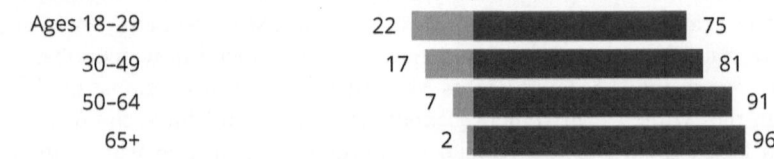

	Phase out	Mix
Ages 18–29	22	75
30–49	17	81
50–64	7	91
65+	2	96

Among Dem|lean Dem who are ...

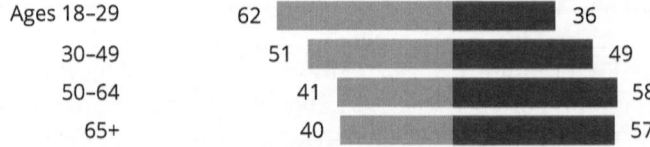

	Phase out	Mix
Ages 18–29	62	36
30–49	51	49
50–64	41	58
65+	40	57

Note: Respondents who did not give an answer are not shown.
Source: Survey conducted May 30-June 4, 2023.

PEW RESEARCH CENTER

Figure 6.7 Americans' attitudes toward fossil fuels. Adapted from a graphic by Pew Research Center.

Many young people, facing a planetary crisis, feel that action is now up to them. On August 14, 2023, a Montana state court decided in favor of young people who filed suit claiming that the state, by promoting fossil fuels, had violated their right to a "clean and healthful environment." The plaintiffs filed this case when they were between ages 2 and 18. One of the plaintiffs was Badge Buss, now age 15. He said, "The fact that kids are taking this action is incredible. But it's sad that it had to come to us. We're the last resort."

Buss's statement was reminiscent of Jerry Garcia of the Grateful Dead who was quoted as saying, "Somebody needs to do something. It's just incredibly pathetic that it has to be us."

> That mix of pride and exasperation [from the plaintiffs] is not uncommon among young climate activists. Many are energized by what they see as the fight of their lives, but also resentful that adults haven't seriously confronted a problem that has been well understood for decades now.[72]

Seventy percent of people aged 16–25 are "extremely worried" or "very worried" about climate, according to a study covering ten countries and published in *The Lancet*.[73] This tells us that we need a genuine conversation between young and old about climate change. Feelings of guilt, fear, and gratitude are part of the picture. Do young people blame the older generation for the world they're inheriting? What sort of action is necessary now? We may not be giving up, but many of us simply don't know what to do or how to do it. What is bubbling up within us that we don't talk about? Can we escape despair by going on a news fast or by selecting facts that support our own preconceived ideas? How do we disrupt people's thinking, including our own thinking? What are the questions we as elders need to hear from young people? What reply can we give?

Perhaps it is not an accident that the oldest President in American history, Joe Biden, succeeded in passing the most far-reaching climate legislation in American history, the Inflation Reduction Act (2022). Ali Zaidi, Biden's deputy national climate adviser, put it bluntly: "Whenever we've achieved a phase change it's been young people making it happen" (Youth Climate Justice Study). Al Gore, Nobel Prize winner for his climate work and now age 75, put it this way: "Morally based movements have gained traction at the very moment when young people decided to make that movement their cause."

Mary Robinson, former President of Ireland and now Chair of The Elders group, said this about climate change: "People would just sort of say 'Ah yes, but that's not me. Having children say 'We have no future' is far more effective. When children say something like that, adults feel very bad." Even the Secretary General of OPEC said this about Big Oil (2019): "Our own children are asking us about their future because . . . they see their peers on the streets campaigning against this industry."[74]

6.2   Mary Robinson: Eco-Elder

Figure 6.8 Mary Robinson. Public domain.

Mary Robinson, born in 1944, was the seventh president of Ireland, serving from 1990 to 1997 and achieving unprecedented popularity. Leaving high office did not mean leaving the world of action. She returned to live in Ireland at the end of 2010 and founded The Mary Robinson Foundation, with attention to Climate Justice. She also took up appointment as United Nations High Commissioner for Human Rights and became an important environmentalist.

Mary Robinson now serves as Chair of The Elders, a group of world leaders in later life striving to make a better world for their descendants. The Elders, founded by Nelson Mandela, have included members such as Ban Ki-Moon (former U.N. Secretary General), Jimmy Carter, Desmond Tutu, and Gro Harlem Brundtland (former Prime Minister of Norway). The Elders are strongly supportive of climate advocacy.[75]

In her role as Chair of the Elders, Mary Robinson has insisted that "only bold, collective action can turn back the hands of the Doomsday Clock" including action on climate. She has said, "The exigency of this situation must not lead us to despair, rather it should propel us into action." She announced her own commitment to eating less meat. Along with her fellow Elder, Kofi Annan, she spoke to 1,300 young leaders from 191 countries encouraging intergenerational action on climate change to take place now. Her message is consistent with HERE NOW YOU HOPE.

Mary Robinson in her later years represents an Eco-Elder on a global scale.[76]

Growing Up and Growing Older

What does it mean to grow up or to grow old at a time of climate crisis? One way to describe it is with the initials VUCA, standing for volatility, uncertainty, complexity, and ambiguity. The climate crisis is more than just extreme weather events. It also describes our situation as one of constant, unpredictable changes in weather patterns all over the globe.

I graduated from college, from Yale, in 1967, at the age of 22, and got married in 1969. The social, economic, and political world that I entered was in turmoil, true enough. The Vietnam War shaped my life at that time. But basic patterns of the world were remarkably stable: e.g., the U.S. Constitution, the cycle of seasons. These remained unchanged. For someone born at the beginning of the 21st century, someone entering the adult world in the decade of the 2020s means that kind of stability cannot be assumed anymore. Instead, young people are entering the adult world at a time of polycrisis.[77] Those who are now retired, people like me, are also experiencing old age at a time not anticipated during our adult life. But the young will face this circumstance for much longer, even if they had less to do with it than others.

Elders, like me, need to ask difficult questions: Do young people blame the older generation for the world they're inheriting? What sort of action is necessary now? True, we may not be giving up in effort. But many of us, young or old, simply don't know what to do about it. What is bubbling up in us elders, and in younger people, what about things that we don't talk about? Can we escape despair by going on a news fast or by selecting facts to support our own preconceived ideas? What are the questions we as elders need to hear from young people? What reply can we give?

Old age is a time when we can no longer take things for granted. We who are old today came of age (post-WWII) in a time the French call *les Trente Glorieuses* (the 30 glorious years of prosperity). Like good health, we took it for granted. So with planet Earth, once completely covered with ice, we took the Holocene for granted, even as the Anthropocene was the new, and still unknown reality before our very eyes. But can we see it? Is denial only for those who reject climate change? Third Act founder Bill McKibben had it right when he said we who are old are closer to the exit than to the entrance. I think about that T-shirt I got for my birthday, the T-shirt that said, "I thought growing old would take longer." I also thought climate change was for future generations. Now I know better, even if I don't know enough.

A survey by the British medical journal *The Lancet* has found that 70% of those aged 16–25 are "extremely worried" or "very worried" about climate. This tells us that we need a genuine conversation between the young and the old about climate change. Feelings of guilt, fear, and gratitude are an essential part of the picture. Nearly 40% said that their fears about the future would make them unwilling to have children of their own.[78] In such a condition of climate anxiety, can we expect younger people to be interested in saving for retirement? In a world where young people see no future, should we be surprised if they blame the old? Consider the words of Greta Thunberg (2019):[79]

> It is we young people and future generations who are going to suffer the most from the climate and ecological crisis. It should not be up to us to take the responsibility, but since the leaders are behaving like children, we have no other choice. The older generations are failing us. And the political leaders are failing us. But we will be watching and holding them accountable.

Abraham Joshua Heschel said: "In a free society, some are guilty, but all are responsible." That means those of us who are older lived through the time when global warming accelerated to what it is today. We can't say we didn't know or that we weren't warned.

Older climate protesters, who spoke with *The Washington Post*, said they felt a sense of collective responsibility, saying, "It was our generation that drove gas guzzlers, thought nothing of flying abroad for a beach vacation in Spain, paid scant attention to deforestation in the Amazon, and, it was on our governments' watch that carbon emissions climbed."[80]

Third Act founder Bill McKibben put it this way: "Our generations have done their share of damage; we're on the verge of leaving the world a worse place than we found it." A protester at an event organized by Extinction Rebellion said this: "I'd do anything to protect my grandchildren," said Charmian Kenner, 67, a retired academic. "I won't live long enough to know whether it worked or not for them, but I'm here, doing this."

When they grow up, if we elders are still here, what will our answer be to their questions: When did you know what was happening to the climate? Those were the questions asked when President Richard Nixon was impeached in 1973: What did you know and when did you know it? Whether or not you have grandchildren, nephews, and nieces or children of others you care about: What did you know and when did you know it?

In response to the question, Who's responsible for global warming? Charmian Kenner (in the photo above) said that older generations "should have acted earlier, so we need to be here, because in some way, whether we meant it or not, we are responsible for what happened." As another protester, John

Lynes, aged 93, a great-grandfather and retired engineer, put it: "Full stop, we are responsible—no doubt about it."

When *NY Times* columnist Margaret Renkl turned 60 years old, she started to think of herself as old and said this:

> "Lately it's been dawning on me that I would not want to have been born even one minute later than 1961, either. Last week I mentioned this new thought to a friend, and her response was immediate, as though she'd already had it herself: 'Because we won't have to live through the cataclysm?'
>
> On the days when headlines are full, yet again, with firestorms, catastrophic flooding, biodiversity collapse, endless pandemic, and a depressingly effective disinformation campaign to deny the climate emergency—on those days, yes. Absolutely yes. On those days I am glad to be 60 because it means I almost certainly won't live to witness the cataclysm that is coming if humanity cannot change its ways in time."[81]

These were Margaret Renkl's words to the younger generation, in a college graduation speech:

> You are children of the 21st century, and yours is the first generation to recognize the inescapable urgency of climate change, the first not to deny the undeniable loss of biodiversity. You have grown up in an age permeated by the noise of a 24-hour news cycle, by needless political polarization, by devastating gun violence, by the isolating effects of "social" media. You have seen hard-won civil rights rolled back. You have come of age at a time of existential threat—to the planet, to democracy, to the arc of the moral universe itself—and none of it is your fault.[82]

What about younger people in the audience, those who will bear the brunt of climate change? Lissa Roy, a 58-year-old grandmother became a member of Extinction Rebellion and joined a public protest where she said: "It saddens me to see young people here, having to do this, risk arrest, knowing the impact that could have on their lives, because we didn't do a good-enough job early on."

What world are we elders leaving to young people? Bella Lack has compiled stories of young people facing the climate crisis.[83]

> This is the planet we're inheriting," [said Bella Lack]. "And we're being given a poisoned chalice, and it's our responsibility to change that. A few days ago my uncle said to me, 'My generation is done. It's up to you now.' And I think that's so damaging. He's only in his 50s.' Turning to me sharply, she says, 'I'm sure you feel responsibility to your kids, to leave them a planet that is the same or better than the one you've lived in.[84]

As in the famous line from *The Death of a Salesman*, "Attention must be paid!" We pay attention by listening to the voices of the younger generation. For instance, Daniel Sherrell (born in 1990) is a member of the Millennial cohort and also the son of a climate scientist and one who became a climate activist himself in young adulthood. His book *Warmth: Coming of Age at the End of Our World* is written as a letter to his own potential future child. In the book he is deeply preoccupied with the impact of climate change on future generations.

Sherrell's own climate anxiety is clear:

The nature of fossil fuel is essentially ambivalent: an invisible pervasion that powers all things and will also, inevitably, destroy them. It is produced by drilling, extracting, and refining those delicate ferns and planktons that millions of years ago were ground to sludge by the planet's crust. The sludge gets processed into coal, gas, or petroleum, then shipped off to power plants where it is converted into energy. In this way we power the present almost exclusively by burning the remains of the past. Unearthed, our history surrounds us, dissolving through the air, until its ubiquity comes to look very much like the future itself.

What will be that future for generations to come? In his book Sherrell shares one of his own climate dreams about catastrophe:

SINKING BENEATH THE WATER

In these dreams I would plunge into a body of water and realize I was sinking. Casually at first, and then with increasing urgency, I would thrash my limbs around, trying my hardest to swim. But no matter how hard I kicked, the surface would continue to recede, and the water would grow blacker and blacker. Eventually it would occur to me that I'd exhausted all my options, and that I was soon going to drown. I would feel my mind wrapping its arms around this act, trying to press it into an alternative formation, squeeze from it some last trickle of hope. And when this failed, the gates of my mouth would open and resignation would come flooding in with the water—a sudden slacking of the muscles, a last-minute shuffling of priorities, the bounds of my foresight snapping back to within arm's reach. Then I would wake up suddenly, and, without moving, stare into the darkness above my face, trying and failing to retain this feeling that was so elusive in waking life: the feeling that I could die, was in fact going to die.[85]

If Sigmund Freud were writing *Civilization and Its Discontents* today, instead of a hundred years ago, then Daniel Sherrell's anxiety would be part of the picture. Freud insisted on the power of the Unconscious. But as one wit put it: The problem with the Unconscious is that it is unconscious. What happens when we become all too conscious of the fear in us? Daniel Sherrell, in his dreams, and when he wakes up suddenly, is an archetypal Millennial, only too well aware of the fear that engulfs him when he thinks about what it means to be "coming of age at the end of our world." Sherrell, as a young man of 30, is also coming of age in an aging society: 10,000 Boomers turn 65 each day, and Sherrell feels strongly that older generations have failed the test of environmental responsibility:[86]

> Rarely do I share these anxieties with any boomers beyond my parents. Their generation did not grow up conscious of the Problem [of Climate Change], and will not live to watch it metastasize. Usually, people over 50 just tell me not to worry about having kids. "It's their generation that's going to solve the Problem!" they'll say, smiling in a way that suggests I lighten up.

> I hate this attitude—how breezily it passes the buck; how it lets itself off the hook and swallows optimism like a sedative. This shouldn't be how it works, each generation abandoning the next to an increasingly impossible situation, waving goodbye and good luck. True intergenerational justice demands more of us.[87]

> To me this was all a cruel irony: how you had to already feel like the world was ending to be able to assimilate the truth when it actually did. And no matter how hard I tried, I couldn't feel it, not consistently. The selective blindness, the congenital optimism—it all felt too hardwired . . . even through my devastation, I felt like I was brushing against some truer reality, the empirical one, the world as it was outside and prior to the filters with which I otherwise convinced myself, silently and as a matter of course, that everything was fine.[88]

Daniel Sherrell's voice is not a statistical sample of climate anxiety in younger generations. But his book, his dream, and his comments can be representative of one answer to the question "Who is to blame for global warming?" We can round up the usual suspects: the fossil fuel companies, the big banks, the rich countries, and the rich people. But at the end of the day, it is children and young people who will inherit the planet left to them by their elders. Even if we say that some are more to be blamed than others, still, as Rabbi Heschel said, all are responsible. Today's elders, including the author of this book, are included in that number, which is exactly why we're obliged to speak out.

There is another saying from the rabbis, this one a Talmudic commentary: "Do not be daunted by world's grief . . . You are not obligated to complete the work, but neither are you free to abandon it." Perhaps the language of "blame" or even "responsibility" is the wrong way to think about how we as elders respond to climate change. Perhaps it is better to speak of "maturity"; this statement is from an American Zen Buddhist priest who tries to offer an answer to the question of what it means to truly grow up, to achieve maturity:

> A mature person is someone who is willing to hear the call, no matter how faint or unexpected it may be, and respond. It is not necessary, however, to look around for things to be responsible for if nothing appears. But when something does appear, you are ready to respond with all your attention and loving care, and with no excuses, no avoidance, no fanfare. You just roll up your sleeves and do it.[89]

Fischer's statement here is akin to what the environmentalist David Orr said about hope: "Hope is a verb with its sleeves rolled up."[90]

The message of age is to use our gifts to outlive the self:

> As we become old, we enter a time of life, even with its losses and deficits, that is not a defective version of youth or middle age, but is something quite different, with its own qualities, discoveries, and surprises . . . This time of life offers new viewpoints in a world that cannot stop its habitual obsession with consuming and polluting . . . It is a time of life we might wish to ignore but which all of us, the living, will and must share. It is inevitable. And it is, in its own way, a gift.[91]

The message of maturity is not one of blame nor allocation of responsibility because we lack the full knowledge for such discernment. Instead, what is required of us is answering the call.

What do we ask of older people? A short answer: a planet for kids like the planet we grew up in.

In the remainder of this, we will look more deeply into what each of us can do to "mend that part of the world that is within our reach." We will look at how we invest our money, how we make consumer choices, and how we act politically as citizens in a world where climate change affects all generations. The ethics of climate action by elders requires us to look unflinchingly at what the task may be, to avoid excuses, and to ask, What Should We Do? which is the focus of the next chapter.

Notes

1 But see also: "Opinion: You Are Not the Problem—Climate Guilt Is a Marketing Strategy" at: https://news.climate.columbia.edu/2023/02/15/you-are-not-the-problem-climate-guilt-is-a-marketing-

2 Barbara H. Fried, "Beyond Blame" Fried Argues That the Philosophy of Personal Responsibility Has Ruined Criminal Justice and Economic Policy and That It's Time to Move Past Blame. *Boston Review* (July/Aug., 2013) at: www.bostonreview.net/forum/barbara-fried-beyond-blame-moral-responsibility-philosophy-law/

3 H.R. Moody, "Elders and Climate Change: No Excuses" *Public Policy & Aging Report* 2017;27(1):22–26 at: https://academic.oup.com/ppar/article/27/1/22/3052917

4 "The Climate Crisis Demands Collective and Individual Leadership and Responsibility" *Common Dreams* at: www.commondreams.org/opinion/climate-crisis-demands-responsibility See also: Britt Wray, *Generation Dread: Finding Purpose in an Age of Climate Crisis* (Knopf-Canada, 2022).

5 Jocelyn Timperley, "Who Is Really to Blame for Climate Change?" *BBC Future* (June 18, 2022) at: www.bbc.com/future/article/20200618-climate-change-who-is-to-blame-and-why-does-it-matter

6 Although I am by background a philosopher and an ethicist, in this book I cannot go deeply into the philosophical questions about justice between generations. For a valuable approach, see Edward Page, *Climate Change, Justice and Future Generations* (Edward Elgar, 2006). See also Elizabeth Cripps, *What Climate Justice Means and Why We Should Care* (Bloomsbury Continuum, 2022).

7 Note, too, that we were all warned about the climate threat a long time ago: (in 1956) www.nytimes.com/interactive/projects/cp/climate/2015-paris-climate-talks/from-the-archives-1956-the-rising-threat-of-carbon-dioxide (in 1965) www.climatefiles.com/climate-change-evidence/presidents-report-atmospher-carbon-dioxide/ See Bill McKibben https://thirdact.org/ and www.theguardian.com/profile/gaia-vince

8 Paula Span, "Older People Are Contributing to Climate Change, and Suffering from It" "The New Old Age" *New York Times* (May 24, 2019).

9 Adam Levy, "Scientists Warned about Climate Change in 1965. Nothing Was Done" *Knowable* (May 30, 2023) at: https://knowablemagazine.org/article/food-environment/2023/scientists-warned-climate-change-1965-podcast See also: "Climate Change Warning Signs Started in the 1800s 'Here's What Humanity Knew and When' Climate Change Concern Has Recently Skyrocketed, But Scientists Began Warning Humanity in the 1800s" at: www.usatoday.com/story/news/nation/2023/06/10/timeline-of-climate-change-what-humanity-knew-and-when/70273996007/

10 Dave Levitan, "Forty Years Ago, EPA Scientists Warned About Climate Change. How Accurate Were Their Predictions?" *The Messenger* (Sept. 1, 2023) at: https://themessenger.com/tech/forty-years-ago-scientists-warned-about-climate-change-how-accurate-were-their-predictions

11 For Hansen's statement see "World Will Look Back at 2023 as Year Humanity Exposed Its Inability to Tackle Climate Crisis, Scientists Say" The Guardian (Dec. 29, 2023) at: www.theguardian.com/environment/2023/dec/29/world-will-look-back-at-2023-as-year-humanity-exposed-its-inability-to-tackle-climate-crisis

12 Post Carbon Institute Power Podcast (Nov. 9, 2022).

13 "Let Them Not Say" at: https://poets.org/poem/let-them-not-say

14 From: "Why So Much Is Going Wrong at the Same Time" *Vox* (Oct. 18, 2023).

15 Robin Wall Kimmerer, *Braiding Sweetgrass: Indigenous Wisdom, Scientific Knowledge and the Teachings of Plants* (Milkweed Editions, 2015).

16 Alice Bell, *Our Biggest Experiment: An Epic History of the Climate Crisis* (Counterpoint, 2021).

17 Cara Buckley, "These Groups Want Disruptive Climate Protests. Oil Heirs Are Funding Them" *NY Times* (Aug. 11, 2022) at: www.nytimes.com/2022/08/10/climate/climate-protesters-paid-activists.html

18 Douglas Starr, "Just 90 Companies Are to Blame for Most Climate Change, This 'Carbon Accountant' Says" *Science* (Aug. 25, 2016) at: www.science.org/content/article/just-90-companies-are-blame-most-climate-change-carbon-accountant-says

19 Charu Sudan Kasturi, "Butterfly Effect; Read to Gulp Down Some Climate Guilt?" *OZY* (Aug. 26, 2021) at: www.ozy.com/around-the-world/butterfly-effect-ready-to-gulp-down-some-climate-guilt/440008/

20 Ajit Niranjan, "Should I Feel Guilty About My Carbon Footprint?" *Made for Minds* (Mar. 1, 2021) at: www.dw.com/en/should-i-feel-guilty-about-my-carbon-footprint/a-60080346

21 Matthew Kotchen, "Offsetting Green Guilt" *Stanford Innovation Review* (Spring, 2009) at: https://ssir.org/articles/entry/offsetting_green_guilt

22 Aled Jones and Bradley Hiller, "Why the Climate Movement Must Do More to Mobilize Older People" *The Conversation* (June 2, 2021) at: https://theconversation.com/why-the-climate-movement-must-do-more-to-mobilise-older-people-161732

23 Robert N. Butler, "Age-Ism: Another Form of Bigotry" *The Gerontologist* 1969; 9(4):243–246.

24 Ashton Applewhite, "How Did Old People Become Political Enemies of the Young?" *The Guardian* (Dec. 23, 2018) at: www.theguardian.com/commentisfree/2018/dec/23/old-people-enemies-young-generational-compact

25 Jenny Womack, "Victims? Villains?—or Possibly Vanguards? The Role of Aging in Climate Change" *Partnerships in Aging* (University of North Carolina, May 21, 2021) at: https://partnershipsinaging.unc.edu/2021/05/victims-villains-or-possibly-vanguards-the-role-of-aging-in-climate-change/ See also: "Older People Are Contributing to Climate Change, and Suffering from It" at: https://desis.osu.edu/seniorthesis/index.php/2021/09/08/older-people-are-contributing-to-climate-change-and-suffering-from-it/

26 H. Estiri and E. Zagheni, "Age Matters: Ageing and Household Energy Demand in the United States" *Energy Research & Social Science* 2019;55:62–70.

27 Concluding lines from Tennyson's poem at: www.poetryfoundation.org/poems/45392/ulysses

28 Dean Baker, "Convincing the Young to Blame the Old, Not the Rich" (Jan. 6, 2016) at: www.commondreams.org/views/2016/01/06/convincing-young-blame-old-not-rich Baker is offering a critique of Catherine Rampell, "College Students Should Aim Their Rage at Older Americans" *Washington Post* (Dec. 24, 2015).

29 "Accountability Is the Most Important Climate Solution" *Drilled News* (Jan. 2, 2023) at: www.drilled.media/accountability-is-a-climate-solution/

30 "Just 100 Companies Responsible for 71% of Global Emissions, Study Says" at: www.theguardian.com/sustainable-business/2017/jul/10/100-fossil-fuel-companies-investors-responsible-71-global-emissions-cdp-study-climate-change

31 "Time to Pay the Piper: Fossil Fuel Companies' Reparations for Climate Damages" *One Earth* (May 19, 2023) at: www.cell.com/one-earth/fulltext/S2590-3322(23)00198-7 See also: "Fossil Fuel Companies Should Pay Trillions in 'Climate Reparations'" *Climate Change News* at: A peer-reviewed paper proposes that the top 21 polluting companies pay $5.4 trillion over 26 years to compensate for climate damages.

32 Al Gore, Ted Talk at: www.ted.com/talks/al_gore_what_the_fossil_fuel_industry_doesn_t_want_you_to_know For more about what Exxon-Mobil knew, see: www.npr.org/2023/09/14/1199570023/exxon-climate-change-fossil-fuels-global-warming-oil-gas

33 If fossil fuel companies are to blame, should they pay reparations? "Fossil Fuel Companies Should Pay Trillions in 'Climate Reparations'" at: https://insideclimatenews.org/news/19052023/fossil-fuel-companies-climate-reparations/

34 Damian Carrington, "Fossil Fuel Industry Gets Subsidies of $ 11M a Minute, IMF Finds" *The Guardian* (Oct. 6, 2021) at: www.theguardian.com/environment/2021/oct/06/fossil-fuel-industry-subsidies-of-11m-dollars-a-minute-imf-finds

35 Kate Yoder, "The True Cost of Climate Pollution? 44% of Corporate Profits" *Grist* (Aug. 28, 2023) at: https://grist.org/economics/true-cost-carbon-pollution-half-of-corporate-profits-climate/

36 "The 4 C's" *Carbon Almanac* (June 12, 2023) at: https://thecarbonalmanac.org/the-4-cs/

37 Kristoffer Tigue, "Concrete Is Worse for the Climate Than Flying. Why Aren't More People Talking About It?" *Inside Climate News* (June 24, 2022) at: https://insideclimatenews.org/news/24062022/concrete-is-worse-for-the-climate-than-flying-why-arent-more-people-talking-about-it/#:~:text=Over%20the%20past%2020%20years,warming%20as%20the%20airline%20industry.

38 Fossil fuel producers are now linked to cement producers: "Fossil Fuel Companies and Cement Manufacturers Could Be to Blame for a More Than a Third of West's Wildfires" at: https://insideclimatenews.org/news/16052023/fossil-fuel-and-cement-companies-blamed-for-third-of-western-wildfires/

39 "Breaking Down Climate Misinformation with Amy Westervelt and John Cook" *Climate One* (Apr. 15, 2022) at: www.climateone.org/audio/breaking-down-climate-misinformation-amy-westervelt-and-john-cook

40 Judd Legum, "Corporations Versus the Climate" *Popular Information* (Aug. 1, 2022) at: https://popular.info/p/corporations-versus-the-climate

41 Jason Anthony, "Aliens on Our Own Planet" *Field Guide to the Anthropocene* (Dec. 1, 2022) at: https://jasonanthony.substack.com/p/aliens-on-our-own-planet?utm_source=substack&utm_medium=email

42 CleanTechnica, "An A to Z Of Fossil Fuel Industry Deception" at: https://cleantechnica.com/2023/05/10/an-a-to-z-of-fossil-fuel-industry-deception/

43 See Astra Taylor, *The Age of Insecurity: Coming Together as Things Fall Apart* (House of Anansi Press, 2023).

44 Nadja Popovich and Brad Plumger, "Who Has The Most Historical Responsibility for Climate Change?" *NY Times* (Nov. 12, 2021) at: www.nytimes.com/interactive/2021/11/12/climate/cop26-emissions-compensation.html See also: "Scientists Have Just Told Us How to Solve the Climate Crisis—Will the World Listen?" at: www.msn.com/en-us/news/opinion/scientists-have-just-told-us-how-to-solve-the-climate-crisis-will-the-world-listen/ar-AAVV7II?ocid=msedgdhp&pc=U531&cvid=3f5a8e4051034b96901d64288978a53b

45 "Per Capita Greenhouse Gas Emissions" https://ourworldindata.org/grapher/per-capita-ghg-emissions

46 Paul Hockenos, "Who Should Foot the Bill for Climate Disasters? Rich Nations, of Course" *CNN Opinion* (Sept. 7, 2022) at: www.cnn.com/2022/09/07/opinions/pakistan-flooding-climate-crisis-pay-liability-us-hockenos/index.html

47 Oliva Rosane, "New NASA Data Shows Which Countries Emit, and Soak Up, the Most Carbon Dioxide" *EcoWatch* (Mar. 14, 2023) at: www.ecowatch.com/nasa-satellite-carbon-dioxide-emissions-countries.html

48 Olivia Rosane, "Wealthy Countries Are More to Blame for Environmental Crises, Study Finds" *EcoWatch* (Apr. 8, 2022) at: www.ecowatch.com/environmental-damage-wealthy-countries.html

49 "Nations Must Increase Funding to Cope with Climate Shocks, U.N. Warns" at: www.nytimes.com/2022/11/03/climate/united-nations-funding-climate-adaptation.html

50 Joseph Winters, "Net-Zero Targets Are More Popular Than Ever, But Less than 5% Are Credible" *Grist* (June 12, 2023) at: https://grist.org/accountability/net-zero-targets-are-more-popular-than-ever-but-less-than-5-are-credible/

51 "John Kerry: Rich Countries Must Respond to Developing World Anger Over Climate" at: www.theguardian.com/environment/2023/jan/03/john-kerry-rich-countries-must-respond-to-developing-world-anger-over-climate?CMP=Share_iOSApp_Other See also: George Monbiot, "There's a Simple Way to Unite Everyone Behind Climate Justice—and It's Within Our Power" at: www.theguardian.com/commentisfree/2022/jun/24/rich-nations-climate-debt-cancelling-debts-emissions-global-debt-swap-campaign?utm_campaign=Weekly%20Briefing&utm_content=20220624&utm_medium=email&utm_source=Revue%20newsletter

52 Jørgen Stig Nørgård, John Peet and Kristín Vala Ragnarsdóttir "The History of the Limits of Growth" at: https://donellameadows.org/archives/the-history-of-the-limits-to-growth/

53 "A Billionaire Is Responsible for a Million Times More Greenhouse Gas Emissions than the Average Person" *Oxfam* (Nov. 6, 2022) at: www.oxfamamerica.org/press/press-releases/a-billionaire-emits-a-million-times-more-greenhouse-gases-than-the-average-person/

54 www.nature.com/articles/s41467-020-16941-y?campaign_id=54&emc=edit_clim_20221118&instance_id=77945&nl=climate-forward&regi_id=20387967&segment_id=113587&te=1&user_id=19f5d92b505a29f7757bf318ae75163e

55 Jared Starr, et al., "Income-based U.S. Household Carbon Footprints (1990–2019) Offer New Insights on Emissions Inequality and Climate Finance" *PLoS Climate* (Aug. 17, 2023) at: https://doi.org/10.1371/journal.pclm.0000190 "How Wealthy 'Super Emitters' Are Disproportionately Driving the Climate Crisis—While Blaming You" The Top 10% of Households Are Responsible for 40% of Total U.S. Greenhouse Gas Emissions at: www.salon.com/2023/08/22/how-wealthy-super-emitters-are-disproportionately-driving-the-climate-while-blaming-you/

56 "Rich People Are the Big Barrier to Stabilizing the Climate" Efforts to curb climate change are failing. That's partly due to the staggering contributions of the global elite: https://newrepublic.com/article/176558/emissions-rich-people-problem

57 Molly Taft, "Billionaires Are Funding Climate Destruction" *Gizmodo* (Nov. 7, 2022) at: https://gizmodo.com/billionaires-are-funding-climate-destruction-1849753810

58 Prarthana Prakash, "125 Billionaires Produce the Same Carbon Footprint as the Entire Country of France" at: https://fortune.com/2022/11/08/billionaires-carbon-emissions-oxfam-report-france/?emci=73566872-9f63-ed11-ade6-14cb65342cd2&emdi=c5d29998-4e64-ed11-ade6–14cb65342cd2&ceid=3972523

59 Joe Fassler, "The Superyachts of Billionaires Are Starting to Look a Lot Like Theft" *NY Times* (Apr. 10, 2023) at: www.nytimes.com/2023/04/10/opinion/superyachts-private-plane-climate-change.html We also need to remember tax breaks: See Paul Kiel, "Private Planes and Luxury Yachts Aren't Just Toys for the Ultrawealthy. They're Also Huge Tax Breaks" *ProPublica* (Apr. 5, 2023) at: www.propublica.org/article/private-jets-yachts-wealthy-tax-deductions-irs-files

60 Emmanuel Saez and Gabriel Zucman, "The Rise of Income and Wealth Inequality in America: Evidence from Distributional Macroeconomic Accounts" *Journal of Economic Perspectives* Fall 2020;34(4) at: https://pubs.aeaweb.org/doi/pdfplus/10.1257/jep.34.4.3#page=8

61 Jack Graham, "As Leaders Fly to Davos, How Do Private Jets Affect the Climate?" *Context* (Jan. 13, 2023) at: www.context.news/climate-risks/as-leaders-fly-to-davos-how-do-private-jets-affect-the-climate?utm_source=pocket-newtab

62 If the bottom 90% had kept up with GDP growth, the vast majority would have brought in $2.5 trillion more in income by 2018. Kathryn A. Edwards and Carter C. Price, "A $2.5 Trillion Question: What If Incomes Grew Like GDP Did?" *RAND Corporation* (Oct. 6, 2020) at: www.rand.org/blog/2020/10/a-25-trillion-question-what-if-incomes-grew-like-gdp.html

63 Somini Sengupta, "The American Exception" *NYTimes* (Feb. 28, 2023) at: www.nytimes.com/2023/02/28/climate/climate-change-carbon-footprint-america.html

64 "How wealthy 'super emitters' are disproportionately driving the climate crisis—while blaming you" The top 10% of households are responsible for 40% of total U.S. greenhouse gas emissions at: www.salon.com/2023/08/22/how-wealthy-super-emitters-are-disproportionately-driving-the-climate-while-blaming-you/

65 See "Tech Billionaires Need to Stop Trying to Make the Science Fiction They Grew Up on Real" at: www.scientificamerican.com/article/tech-billionaires-need-

to-stop-trying-to-make-the-science-fiction-they-grew-up-on-real/ See also: "The Acronym Behind Our Wildest AI Dreams and Nightmares" at: www.truthdig. com/articles/the-acronym-behind-our-wildest-ai-dreams-and-nightmares/

66 Douglas Rushkoff, *Survival of the Richest: Escape Fantasies of the Tech Billionaires* (Norton, 2022).

67 See Thomas Piketty, *Capital in the Twenty-First Century* (Harvard University Press, 2017); and Thomas Piketty, *A Brief History of Equality* (Harvard University Press, 2022). See also Joseph Stiglitz, *Great Divide: Unequal Societies and What We Can Do About Them* (W.W. Norton, 2015).

68 Keerti Gopal, "New York Activists Descend on the Hamptons to Protest the Super Rich Fueling the Climate Crisis" *Inside Climate News* (Aug. 5, 2023) at: https://inside-climatenews.org/news/05082023/hamptons-protest-super-rich-climate-crisis/

69 Matthew Huber, *Climate Change as Class War* (Verso, 2022).

70 Richard Schiffman, "Gen Z, Climate Change Is a Heavy Emotional Burden" *Yale Environment 360* (Apr. 28, 2022) at: https://e360.yale.edu/features/for-gen-z-climate-change-is-a-heavy-emotional-burden

71 Pew Research Center, "What the Data Says about Americans' Views of Climate Change" (Apr. 18, 2023).

72 David Gelles, "With TikTok and Lawsuits, Gen Z Takes on Climate Change" *New York Times* (Aug. 19, 2023) at: www.nytimes.com/2023/08/19/climate/young-climate-activists.html

73 From Katharina Buchholz, "This Chart Shows Global Youth Perspectives on Climate Change" *Statistica, World Economic Forum* (Oct. 26, 2022) at: www.weforum. org/agenda/2022/10/chart-shows-global-youth-perspectives-on-climate-change/

74 https://youthclimatejusticestudy.org/why-youth-why-now-2/

75 For The Elders, see: https://theelders.org/programmes/climate-change

76 For more on Robinson's concern for future generations see: "Our responsibilities to the future of the world" at: https://theelders.org/news/2023-and-beyond-our-responsibilities-future-world

77 Daniel Hoyer, et al., "Navigating Polycrisis: Long-Run Socio-Cultural Factors Shape Response to Changing Climate" *SocArXiv Papers | Navigating Polycrisis: Long-Run Socio-Cultural Factors Shape Response to Changing Climate (osf.io)* (Mar. 31, 2023) http://doi.org/10.1098/rstb.2022.0402 at: https://osf.io/preprints/socarxiv/h6kma/

78 Schiffman, "Gen Z, Climate Change Is a Heavy Emotional Burden."

79 Greta Thunberg, "The Older Generations Are Failing Us" *StreetsBlog Denver* (Oct. 11, 2019) at: https://denver.streetsblog.org/2019/10/11/greta-thunberg-in-denver-the-older-generations-are-failing-us/

80 Karl Adam, " 'Gray Greens': Grandparents Are Being Arrested in London Climate Protests" *Washington Post* (Sept. 4, 2021) at: www.washingtonpost.com/world/europe/climate-protests-london-arrests/2021/09/04/8e6cf6be-0bf1-11ec-a7c8-61bb7b3bf628_story.html

81 Margaret Renkl, "I Just Turned 60, But I Still Feel 22" *New York Times* (Nov. 1, 2021) at: www.nytimes.com/2021/11/01/opinion/turning-60-aging.html

82 Margaret Renkl, "My Generation Wrecked So Much That's Precious. How Can I Offer You Advice?" *NY Times* (May 15, 2023) From her Baccalaureate Address to the University of the South at: www.nytimes.com/2023/05/15/opinion/letter-to-graduates-hope-despair.html

83 Bella Lack, *The Children of the Anthropocene: Stories from Young People at the Heart of the Climate Crisis* (Penguin, 2022).

84 Bella Lack, from an interview at: www.theguardian.com/environment/2022/aug/14/whats-the-alternative-to-give-up-environmentalist-bella-lack-the-new-queen-of-green?CMP=Share_iOSApp_Other

85 Daniel Sherrell, *Warmth: Coming of Age at the End of Our World* (Penguin, 2021), pp. 25–26.

86 Is it "Boomer Bashing" to blame those born between 1946 and 1962? We hear that sentiment often: "It's baby boomers' world. Millennials and Gen Z'ers fear for their financial future because they're still living in it." We don't need to agree with this article. But we do need to pay attention: https://fortune.com/2023/09/16/millennials-gen-z-worry-baby-boomer-impact-financial-future/

87 Sherrell, *Warmth*, p. 249.

88 Ibid., p. 26.

89 Norman Fischer, *Taking Our Places: The Buddhist Path to Truly Growing Up* (HarperSanFrancisco, 2003).

90 David Orr, *The Essential David Orr* (Island Press, 2010).

91 From Douglas Penick, "A Few Thoughts on This Admission: I Am Old" *Common Dreams* at: www.commondreams.org/opinion/our-old-age

7   What Should We Do? No Time for Excuses[1]

The most important thing every single one of us can do about climate change is talk about it.

—Katharine Heyhoe, climate scientist

There's an old joke asking why is it that the American people don't act in response to a big challenge, such as climate change: The question: Are people ignorant or apathetic? The answer: "I don't know and I don't care."

Joking aside, there is a serious question[2] about why people resist truths about climate change.[3]

Alas, people of all ages, including elders, are among those who resist the threat of climate change. The massive threat of climate change has been described as the greatest moral challenge of our age, with specific reference to the idea of legacy.[4] Reb Zalman Schachter said to me, the question is: Are you doing your legacy work? Our answer is, evidently not. Environmentalism has clearly been identified as an issue for an aging population,[5] yet, until recently, it was hardly ever discussed at meetings of groups advocating on behalf of older people. Is that because they are ignorant or apathetic?

Among gerontologists there has been some recent welcome attention to aging and climate change.[6] The American Society on Aging devoted a special issue of its journal to the subject of climate change[7] Some have called for an active research agenda on sustainability and population aging.[8] A key element in that research agenda will be better understanding of communication about climate change in an aging society.[9] Researchers on climate change communication have noted:

> Nearly all climate scientists are convinced that human-caused climate change is occurring, yet half of Americans do not know or do not believe that a scientific consensus has been reached. That such a large proportion of Americans do not understand that there is a near-unanimous scientific consensus about the basic facts of climate change matters.[10]

There is growing recognition among climate scientists and environmentalists that their communication about climate change has, somehow, missed

DOI: 10.4324/9781003345992-10

the mark. For example, Peter Tague, a leading environmental philanthropist, put it bluntly: "Over the last 15 years environmental foundations and organizations have invested hundreds of millions of dollars into combating global warming. We have strikingly little to show for it." What is the reason?

This is no less true for older people than it is for the general population. A 2018 Gallup poll showed evidence for what they called a "global warming age gap." Seventy percent of adults aged 18 to 34 expressed worry about global warming compared to 56% among those over 55. A later analysis confirms the main point but shows a more complex picture.[11] In an effort to provide some response, it is useful to start by accepting that older people themselves have sincere, and serious, objections to actions on climate change. Instead of dismissing these objections, here we want to listen to some hypothetical objections and then offer reasons that can also be taken seriously. One of my premier mentors. Sam Sadin, taught me this about marketing: The sale begins only when the customer says no. So let's look at some of the ways that people say "no" to acting on climate change. These are the objections that need to be answered. This is the point being made by Katharine Heyhoe when she tells us that the most important thing we can do about climate change is to talk about it. No, it's no good to reply that "talk is cheap" and that talking about a subject is "just words." To talk about climate change means to listen honestly to objections and then offer a reply that responds to that objection. Here are some objections to be considered.

Uncertainty About the Future: Who Knows What Will Happen?

Here is the first objection to be considered:

> The future is always uncertain. We've heard scare stories before We older people remember that "Y2K" bug, with predictions of worldwide computer collapse in the year 2000. We remember the Swine Flu epidemic that never happened, not to mention SARS, Swine Flu, bird flu, and all other disasters warned about by "medical experts." Some of us even remember a *Time* magazine cover story (1975) about a "coming Ice Age" (global cooling, not warming), not to mention environmentalist Paul Ehrlich's warning of Famine 1975! These things never happened. Now here come the global warming experts, with their "models" forecasting more bad news of things to come. No one can know the future.

Older people are natural skeptics, and long life experience gives good grounds for their skepticism. Many predicted disasters never happened, so is it any wonder that today's seniors have doubts about climate change? But the fact that some predictions didn't come true doesn't mean that all predictions are wrong. We don't usually ignore weather reports—about rain tomorrow—even though weather reports aren't always accurate.

But is uncertainty, at any level, a reason for not acting? But we don't normally live like that. We take account of the level of uncertainty and the level of a threat. In the case of climate science, there is overwhelming, nearly unanimous agreement that it's a reality. As we've seen in the first four chapters of this book—the Four Horsemen of the Climate Apocalypse—the threat is already here. It is NOT something for the future. Most of the uncertainty about climate change is uncertainty about timing or details such as the role of clouds reflecting sunlight. But the basic science is clear, despite how vested interests try to confuse matters. Those promoting doubt today are acting the same way that tobacco companies did for many years when confronted with the facts about lung cancer. They tried to promote doubt when mainstream science recognized clear patterns of evidence.

In any case, the "merchants of doubt" are eventually overturned by the facts. A poll by the Pew Research Center in 2020 revealed that more than two-thirds of Americans now recognize that the planet is getting warmer.[12] Denial about climate change is quickly going away when even the big fossil fuel companies no longer engage in direct denial: instead of denial, they favor another "D" word: delay. And it's not just fossil fuel companies who've turned their eyes away: "Twenty-five years ago people could be excused for not knowing much, or doing much, about climate change. Today we have no excuse," said Desmond Tutu, Former Archbishop of Cape Town and winner of the Nobel Prize.

Ok, It's a Problem, but We Have Time

Don't we have time on climate change? It sounds like action, but it's just the other "D" word: delay. Compare what climate scientists say to a biopsy report from the lab that we don't bother to pick up. Maybe I have a problem, but I have time. But when we get the report and read it, our mind begins to change: "We all think we have time, until we don't. It's amazing to me how different our outlook can become when a medical condition threatens to shorten the number of years ahead," said Braga Meininger.[13] Some people change their minds about global warming when extreme weather events happen. Others just go back to sleep. Once we realize that the climate threat is real, we may also realize that we don't know how much time we have at all.

Uncertainty About Science

OK, maybe the climate is changing. But scientists don't really understand why it's happening or what should be done about it, if anything. How can we act if we don't understand how it works?

Uncertainty is always part of science. We never get complete certainty, as older people will remember since they lived long enough to see science change: for example, for years medicine had the wrong explanation for ulcers.

Is there room for skepticism and uncertainty about climate change? Yes, predicting the future of science always involves uncertainty. For example, climate scientists are continuing to investigate the role of different factors, such as clouds and ocean currents, affecting climate change. But the pattern of global warming is clear from measured data, even where we can't be sure about all the details. True, we can't be certain about just how much the earth will heat up with specific concentrations of carbon dioxide. Yet we have known for a hundred years that carbon dioxide produces the greenhouse effect. Nobel-prize-winning chemist Svente Arrhenius (d. 1927) called attention to it over a century ago.[14] The basic pattern is not uncertain at all. Heating may be more or less, but the overall direction is clear. There is legitimate uncertainty about what impact climate change will have on human society or specific geographic areas: for example, how far will coastal flooding affect major cities like New York or Miami.

All these questions call for more research. But we can't wait for final answers before reducing carbon pollution, which is the main trigger for climate change.

Talking About Things Doesn't Make Any Difference in the Real World

> There's lot of talk today about global warming. But talk is cheap. It fills the pages of publications, like the news about fires and crimes: "If it bleeds, it leads" is a slogan among journalists. It's all just talk anyway. I ignore most of what I see in the papers or what's on the news.

Older people are very familiar with the style of mass media. The media does try to emphasize the negative (crimes and fears), and the media does like to play up controversy. That's why we read stories that are framed in terms of "opposing sides," as if both sides were equally plausible. We don't apply that approach to certain controversies: for example, did the Holocaust really happen? There will always be different opinions. What should count is the weight of opinion among well-informed experts. When it comes to climate change, it turns out that 98% of informed climate scientists agree that it's happening.

"Something Will Turn Up" and "Science Will Save Us"

> People have been predicting disasters since Thomas Malthus warned that population growth would result in famine. But, somehow, technology always seems to come up with a solution, like the Green Revolution. Why shouldn't that happen with climate change?

The belief that "something will turn up" is what we might term "magical thinking." We can always hope that science will find a way to help us cope with climate change. People are already working on this: for example, renewable energy costs have come down by 80% or more in the past decade or two.

Could carbon capture be the next technology that will remove greenhouse emissions from the air? A technological breakthrough could happen. But will it happen quickly enough? We can't count on those solutions being available in a timely way even if we support more research and development in these areas, as the 2022 Inflation Reduction did support them. Environmentalists should always be in favor of new technologies that could help solve the problems.[15] So, yes, science and technology could help address climate change, but only if we take the threat seriously and support innovations to address the genuine problem. But the challenge is time: Will we have the technologies available, at cost and at scale, in the years ahead?

Anyway, What Can One Person Do?

I've lived long enough to know that one person alone can't change the world. These problems are just too big. I have enough to do every day as it is.

We could apply this same logic to being a member of a firing squad. Any member of the firing squad could say, "OK, I'll shoot, what difference does one person make?" One person, alone, may not be the total cause of any outcome for any event. But individuals, taken together, do make a difference. Older people have had enough experiences in their lives to remember when the act of a single person turned out to change other people's minds and make a difference in the outcome.

If we take this objection seriously, we can respond by asking what "one person" can do at the level where they can make a difference: for example, acting locally with others concerned about the global problem. Or acting on specific issues—such as waste management or energy efficiency—which do have a big impact at the local level. There's a good reason why environmentalists urge us to "Think global, act locally." Others have put together practical steps for mitigation of climate change and coping with what will happen even during the lives of older people today.[16]

Why Should Rich Countries Curb Emissions When Other Countries Do Nothing?

All those international conferences on climate change have been a failure: Rio, Copenhagen, Paris, and all the rest. We could curb our carbon output in America, but others won't follow, or else they'll cheat. The USA always ends up being a sucker, as you see with foreign aid. A lot of it is just wasted. If we were to cut back on carbon emissions, it wouldn't make up for what China or India are doing. They're opening new coal-fired power plants every day.

There's a valid point here and it was developed in the preceding chapter. Rich countries, along with rich people, have failed to respond to the question of "Who's to blame for global warming?" International diplomacy on the environment has been disappointing. But there are notable exceptions: for example, the Montreal Protocol (1985) banned aerosol to protect the earth's ozone layer. Most observers believe that if the USA acted more decisively on climate change, other countries would be encouraged to follow. China has begun revising its policies and is an active exporter of tools for solar energy. China and the USA have entered into a bilateral agreement to reduce carbon emissions. Germany has dramatically reduced fossil fuel consumption and is a leader in solar power.

The USA should not be falling behind other countries. It's not true that "other countries do nothing." In recognition of this reality, the USA has now rejoined the Paris Accord on climate change. But the USA cannot lead internationally unless it acts domestically to work against climate change, as happened with the Inflation Reduction Act in 2022. In a later chapter, on "Citizen Climate Action," we will look more deeply at the politics of climate change in an aging society.

It's Too Late: We Just Have to Cope as Best We Can

> The momentum for climate change is too great. We've been putting carbon dioxide into the atmosphere for so long that it's too late to stop climate change.

This objection is the polar opposite of "The future is uncertain." Instead, the objection is that everything is certain; we hear the objection that climate change is real but it's hopeless. It's all over and done with: it's just too late. This objection makes a valid point because climate scientists themselves can't predict just how quickly climate change will take place. Measurements show that carbon dioxide in the atmosphere is way past the level of 350 parts per million that could be considered safe. But this point is not an argument for doing nothing. It's an argument that makes it more important than ever to do what is possible to prevent further damage to the earth. "Coping as Best We Can" could involve what is termed "mitigation" and it could also mean "adaptation" to what climate change is already doing: for example, preparing for fires, floods, and other threats of the Climate Apocalypse. None of this can happen as long as people say "it's too late," as Rebecca Solnit and others have reminded us[17]

All This Fear About Climate Change Is Just Extremism, Just Hysteria About Things That Are Unthinkable, Right?

We need to think about what is unthinkable. Before 2016, it was unthinkable that Donald Trump would become President of the USA. Before February 2022, it was unthinkable that Russia would launch a full-scale land war in Europe. We thought it unthinkable that the Arctic ice cap could completely melt in our lifetime. Today there are many who regard it as unthinkable that modern industrialized societies would turn the entire ocean into a dumping ground that is heating to the point where weather patterns themselves could change.

To imagine something is unthinkable is a temptation for complacency, just as it is to imagine that someone else will take the steps needed to stop climate change and preserve our human life on planet Earth. The great danger today is not denial but delay and complacency. These are laziness and fairy stories that keep us in a condition of learned helplessness.

I'm a Person of Faith: It's All in God's Hands Anyway

Whatever happens is in God's hands. That's what I've believed all my life. There have always been plagues and disasters in the world, as early Christians could see during the decline of the Roman Empire. People of faith should have faith in the Divine Plan.

In the past decade larger and larger numbers of Christian Evangelicals have come to believe that failing to curb climate change is a deep and sinful offense against our stewardship for divine creation.[18] Faith in a Divine Plan doesn't stop most people from seeking medical treatment. People of faith can argue that the illness is part of the Divine Plan, but so is the cure. Religious groups are increasingly paying attention to the virtues of stewardship, which is also a Biblical injunction. Most of us don't appeal to "God's plan" when it comes to personal illness. Why should we apply that argument to climate change?

This Problem Doesn't Affect Older People. It's for the Future

OK, maybe there's a problem. But I feel fine now and, in any case, people of my age won't be much affected by climate change. I won't live that long. I'm old now and I'm glad I won't be alive in the future to see it.

This was exactly the response from the woman who heard my speech about climate change: "I'm glad I won't live that long." But sticking one's head in the sand doesn't make problems go away. The Four Horsemen of the Climate Apocalypse are already here: fire, flood, drought, and heat wave. People who are old today have, over the course of their lives, contributed much more to creating the problem of climate change than, say, newborns or children. Yet it is our children who will bear the burden of dealing with a problem that elders have created. That disparity of generational responsibility is a huge ethical problem which must be addressed. Bill McKibben, founder of Third Act, praised the work of young people in climate advocacy, but he also said: "So far the kids have had to do all of the work and they've done an amazing job, but it's not fair to ask 18-year-olds to solve this problem." People of different age groups have different views about action around climate change:

As for seniors who believe they're not yet being affected by climate change, this belief isn't supported by the facts, as we've seen in the chapters of this book documenting the threats: fire, flood, drought, and heat wave.[19] For example, three-quarters of the deaths that happened in Hurricane Katrina were people over 65. Climate scientists predict that hurricanes in the future,

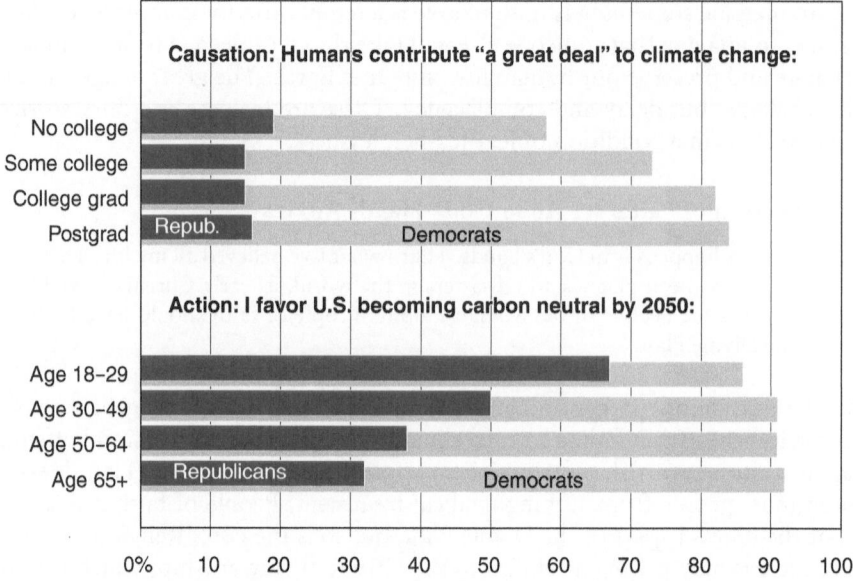

Figure 7.1 Americans' attitudes toward causes of climate change and the USA becoming carbon neutral by 2050 based on Pew Research Center data. Public domain.

along with coastal flooding, are likely to be more severe and elders are in the bull's eye of this threat.[20] Advanced age represents a significant risk for heat-related deaths in the USA, and we can expect that the scale of this problem will increase only as the U.S. population over 65 goes from 13% to nearly 20% by2030. In the near future, climate change will become an immediate problem for older people. Protecting seniors from environmental disaster is increasingly recognized as a problem for today, not just for the future.[21]

People who just don't care about anything that happens after they're dead are very, very few in numbers, as the life insurance industry understands well. The importance of generativity as individual motivation needs to be connected to our collective response for future generations.[22] When we speak to people about their children or grandchildren, we're likely to get a more receptive response. The question we should be asking is, What kind of legacy are leaving to our children and grandchildren?[23]

Climate Change Is a Problem for the Young People to Solve

OK, there's a problem. Young people will solve it. We seniors can leave it up to them.

Bill McKibben responded this way: "I've heard one too many people say to me, 'Oh, it's up to the next generation to solve these problems,' which seems ignoble but also impractical."[24]

It's impractical for political reasons. Older voters are a major voting bloc that can influence elections and public policy. At a minimum, seniors need to be supporting candidates and policies reflecting the threats we are facing. If we're going to "leave it up to the young people," then perhaps older people should work in collaboration with the young to address the problems. If it's a problem for younger people to solve, then intergenerational action is what is called for. Intergenerational cooperation is very different from "leaving it for young people to solve." If this is "a problem for the young people to solve," it also requires that elders be part of the solution, too, which is an argument for collaboration across generations.[25]

7.1 James Hansen: Eco-Elder

James Hansen, Ph.D., born in 1941, spent his career as a planetary scientist and served as Director of NASA's Goddard Institute for Space Studies. He already had a distinguished scientific career when, in 1988, he testified before Congress, saying that "Global warming has reached a level such that we can ascribe with a high degree of confidence a cause and effect relationship between the greenhouse effect and observed warming . . . It is already happening now." His testimony was what historian Douglas Brinkley called "the opening salvo of the age of climate change."

Hansen has spoken out vigorously against the coal industry and against mountaintop removal for that purpose, and he also opposed the Keystone Pipeline. In 2006. In contrast to many academics, even within climate science, Hansen's career has been notable for speaking truth to power. He continued in that advocacy role, both as a government official and after retirement from government in 2013, when he became a bona fide elder.

His biggest concern is for the welfare of future generations, as seen in his book *Storms of My Grandchildren: The Truth About the Coming Climate Catastrophe* (Bloomsbury, 2010). In 2006 *Time Magazine* designated Hansen as one of the 100 most influential people on the earth. In Hansen's own words: "The greatest danger hanging over our children and grandchildren is initiation of changes that will be irreversible on any time scale that humans can imagine . . . What we are doing to the future of our children, and the other species on the planet, is a clear moral issue."

In old age, Hansen has continued his willingness to speak out in controversial ways about climate change, even if it means challenging established sources like the U.N. Intergovernmental Panel on Climate Change. By 2023 Hansen warned that a decrease in aerosol pollution could drive temperatures higher.[26]

How Can You Ever Really Care About Remote Generations?

When we read about climate change, it seems like it might affect remote generations—people living hundreds of years from now. We can't know much about
those people or feel connected to them. Anyway, how can we have responsibilities to people who don't yet exist?

The Founding Fathers of the United States could not imagine the world that
we're living in today. Yet they wrote the U.S. Constitution with an eye to
what remote posterity would need and they did a pretty good job at that.
Moreover, the Founding Fathers were intensely concerned with posterity.
They didn't adopt the mindless quip, attributed to Groucho Marx, "What
has posterity ever done for me?"

It's true that we don't know much about humans on the earth a hundred
years from now. But we can recognize that they will need a safe and predictable environment like the one elders have grown up in or the climate that the
human species has enjoyed for 10,000 years. Our actions today are putting
that environment at risk. Failing to provide a decent world for our descendants is not responsive to justice between generations.[27]

It is a mistake to think that climate change is an issue for remote future
generations. Climate change has already produced increased forest fires in
the USA and rising ocean levels will be a factor in future hurricanes, as we've
seen from Hurricane Katrina and Hurricane Sandy. It's estimated that by

Figure 7.2 Four generations of Americans (2008). Public domain.

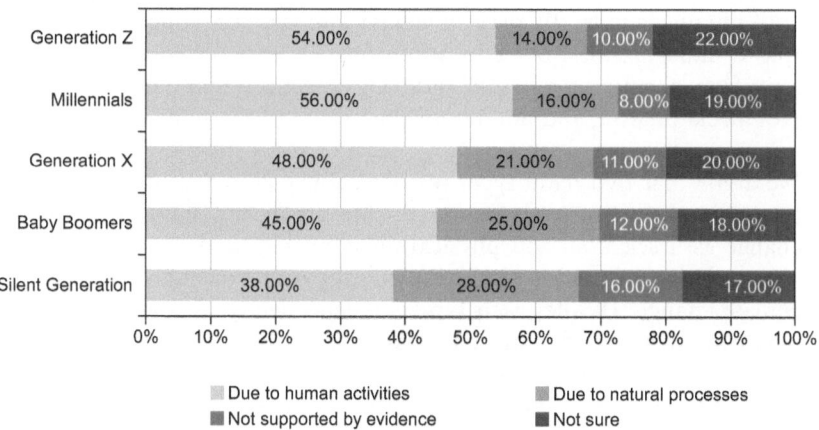

Figure 7.3 Americans' views on causes of climate change based on Pew Research Center data. Public domain.

mid-century—when today's children are grown up—between $66 billion and $106 billion worth of property in the USA will be below the sea level. And that doesn't begin to count the cost of disaster relief shared by all Americans. We're not talking about responsibilities to "people who don't yet exist." We're talking about responsibilities to our children and grandchildren.

Don't Think About It

> This is all just so depressing. It's really hopeless and old age has enough problems. I hear all this stuff and then I try not to think about it.

People can try not to think about it, but climate change is already part of the future of an aging society.[28] It is not going away. People who "don't get it" or who are ignorant about climate change are susceptible to denial: that is, putting their heads in the sand. People who do "get it" are susceptible to despair. For both cases, denial and despair, the result is paralysis and failure to act.[29] Either "What Me Worry?" or "Forget about it" are poles of this paralysis. Erik Erikson described the psychological challenge of old age as the polarity of "Ego-Integrity versus Despair."

Other common beliefs about action on climate change are economic objections, such as "A carbon tax would destroy jobs," "The cost of reducing CO_2 emissions would be prohibitively high," or "Slowing the pace of climate change would be prohibitively difficult." All of these objections can be disputed as "myths," but they depend on complex economic calculations and are likely to be less influential for elders than the objections considered here.[30]

It is important to "inoculate" elders against disabling doubts, doubts that are promulgated by vested interests and wealthy interest groups who want to create uncertainty to prevent action on climate change.[31] It is a failure tactic, and eventually it fails. To prevent denial and despair, education for older people is indispensable. But education must be carefully designed in ways that confront skepticism and take seriously objections that older people have.

Not all skepticism, and not all objections, are of the same order:[32]

> We argue that two main types [of objections] should be distinguished: epistemic skepticism, relating to doubts about the status of climate change as a scientific and physical phenomenon; and response skepticism, relating to doubts about the efficacy of action taken to address climate change. [While] each type is independently associated by people themselves with climate change skepticism, we find that the latter is more strongly associated with a lack of concern about climate change. As such, additional effort should be directed towards addressing and engaging with people's doubts concerning attempts to address climate change.

Research on objections by older people to climate action will have to take account of serious and sincere objections, such as those formulated in this discussion.

What Is to Be Done?

Once we respond to all the objections, the question becomes, What is to be done? In later life, the great philosopher Immanuel Kant said that all of philosophy could be boiled down to just three questions: What do I know? What should I do? What may I hope?

We can apply Kant's three questions to elder's objections to action on climate change. The first step is a response to the question, "What do I know?" Older people need to educate themselves about environmental threats and what can be done. There is no substitute for genuine knowledge. The second step is a response to the question, "What should I do?" The answer is to take action, on whatever scale is possible: for example, to live in a more environmentally responsible fashion and to connect with others who care about threats to our environment. We act more effectively when we work together with others. The third step is a response to the question, "What may I hope?" Here we need to avoid extreme or hysterical thinking: for example, "It's too late" or "Nothing can be done."

Consumerism and mindless materialism are examples of Thoreau's dictum that "most men live lives of quiet desperation." Yet elders themselves typically show a decline in consumerism, and older people can actually be major contributors to society. Desperation is not our only choice. Paul Hawken,[33] in his hopeful book *Blessed Unrest*, points to success stories of citizen action around the globe. We need to celebrate our successes, even as we recognize the severity of environmental threats. What may we hope? is exactly the key question to ask.

Part of the response of hope is to see population aging itself as a path to hope, not a reason to be gloomy, as Robert Butler[34] eloquently reminded us in *The Longevity Revolution*. Population aging, now a global phenomenon, is very good news. It means that the world population can stabilize and eventually decline, making all environmental problems more manageable. What is called for today is for people to overcome objections to action on climate change and to rely on clear evidence for timely action. This can only happen when all generations realize the future is in our hands.[35] Jonas Salk reminded us that "our greatest responsibility is to be good ancestors."[36] It is time for elders to be the good ancestors we are capable of being.[37] Or, as Ashton Applewhite put it, it to "act like an ancestor."[38]

Becoming a good ancestor is also a matter of timing, and it demands a two-fold response, like the Roman god Janus who looks in two directions at once. On the one hand, we don't have time, because time is short. That means there is no time to delay acting on climate change. On the other hand, it's not too late to act. We need both fear and hope.[39] Today is not a time for

Figure 7.4 Marble bust of the Roman god Janus. Public domain.

despair but for action. A Chinese proverb puts it best: the best time to plant a tree was 20 years ago. The second best time is today.

Some guidance for what to do:

- Do not succumb to the tempting anesthesia of complacency or cynicism. The stakes are too high. Even if you cannot take much time out of your normal life for direct action, you can organize, mobilize, and energize friends, colleagues, and neighbors.
- Counter lies with truth. When someone repeats a lie about climate change, correct it. This requires preparing with facts, logic, analysis, and sources.
- Don't waste time commiserating with people who already agree with you. Don't gripe, whine, wring your hands, or kvetch about how terrible is the climate threat. Don't criticize Biden and the Democrats for failing to communicate more effectively how what they're doing about climate advocacy.
- Don't get distracted by the latest sensationalist post or story about extreme weather events. The media has a short-term attention span to divert our eyes from the real prize—the survival of humanity and democracy, during a time when climate action is threatened in one of the greatest stress tests in American history, a stress organized by one of the worst demagogues in American history.[40]

7.2   Lewis Mumford: Eco-Elder

In 1975, I was barely 30 years old and had just started in gerontology at the Brookdale Center on Aging of Hunter College, where I was the sole staff member facing an uncertain future. I was also teaching a course, at New York University, on "The Crisis of Civilization," where I expounded on thinkers like Jacques Ellul and Arnold Toynbee. I was also deeply interested in the environmentalist Lewis Mumford, whose two-volume *Myth of the Machine* had appeared just a few years before. Mumford was then 79 years old and had no idea who I was, but I wrote him a personal letter. He replied to my letter and invited me to visit him at his home upstate in the charming town of Amenia, New York.

I did visit and spent an entire day listening to the story of his life, while given a tour of his beautiful garden. The man himself was all I hoped he would be: a source of inspiration for the course I was teaching, as shown in Mumford's own words:

> Modern Man is the victim of the very instruments he values most. Every gain in power, every mastery of natural forces, every scientific addition to knowledge, has proved potentially dangerous, because it has not been accompanied by equal gains in self-understanding and self-discipline.

Why was Lewis Mumford willing to meet an unknown person a half-century younger than him? Perhaps the reason is found in his statement expressing an understanding of the real ties between generations: "Every generation revolts against its fathers and makes friends with its grandfathers." As part of the generation of the sixties myself, I was looking for a grandfather figure, and Mumford fit the bill. Despite being of a younger generation, I was not comfortable with the eco-pessimism evident in *The Limits of Growth* (1973), published just a few years earlier and a book I treated in my crisis of civilization course. The book was sponsored by the Club of Rome, whose president, Alexander King, was an Eco-Elder who I would come to meet years later.

Lewis Mumford's hope for the future was an abiding gift to all generations. Mumford would remain a productive thinker and a critic of society during his old age, continuing to warn of threats from our technological civilization. He would die, at age 94, many years after our meeting. Mumford once said, "I would die happy if I knew that on my tombstone could be written these words, 'This man was an absolute fool. None of the disastrous things that he reluctantly predicted ever came to pass!'" Mumford's message of his hope is one I have never forgotten.

Notes

1 Picture from Third Act.
2 Mike Hulme, *Why We Disagree About Climate Change: Understanding Controversy, Inaction and Opportunity* (Cambridge University Press, 2009).
3 Clive Hamilton, *Requiem for a Species: Why We Resist the Truth About Climate Change* (Routledge, 2010).
4 Howard Frumkin, Linda Fried and R. Moody, "Aging, Climate Change, and Legacy Thinking and Legacy Thinking" *American Journal of Public Health* Aug. 2012;102(8):1434–1438.
5 Peter Heller, *Who Will Pay?: Coping With Aging Societies, Climate Change, and Other Long-Term Fiscal Challenges* (International Monetary Fund, 2003).
6 For an early contribution about aging and climate change, see S.D. Wright and D. Lund, "Gray and Green?: Stewardship and Sustainability in an Aging Society" *Journal of Aging Studies* 2000;14(3):229–249.
7 See the Summer, 2022 edition of the journal *Generations*, from the American Society on Aging, a collection edited by Mick Smyer: "The Climate Crisis: What's Aging Got to Do With It?" at: https://generations.asaging.org/summer-2022
8 K. Pillemer, N.M. Wells, L.P. Wagenet, R.H. Meador and J.T. Parise, "Environmental Sustainability in an Aging Society: A Research Agenda" *Journal of Aging and Health* 2010;23(3):433–453. Pillemer and colleagues at Cornell have launched an Aging and Climate Clearinghouse, and I edit a weekly newsletter "Climate Change in an Aging Society."
9 Tanya Tillett, "Climate Change and Elderly Americans: Examining Adaptability in an Aging Population" *Environmental Health Perspectives*, Jan. 2013;121(1):a33. Published online Jan 1, 2013. http://doi.org/10.1289/ehp.121-a33 at: www.ncbi.nlm.nih.gov/pmc/articles/PMC3553457/

10 E. Maibach, T. Myers and A. Leiserowitz, "Climate Scientists Need to Set the Record Straight: There Is a Scientific Consensus That Human-Caused Climate Change Is Happening" *Earth's Future*, Wiley Online Library, May 2014 at: http://onlinelibrary.wiley.com/doi/10.1002/2013EF000226/full

11 Matthew Ballew, et al., "Young Adults, Across Party Lines, Are More Willing to Take Climate Action" *Yale Program on Climate Change Communication*, Climate Note, 2020 at: https://climatecommunication.yale.edu/publications/young-adults-climate-activism/

12 Pew Memorial Trust, "Two-Thirds of Americans Think Government Should Do More on Climate" (June 23, 2020) at: www.pewresearch.org/science/2020/06/23/two-thirds-of-americans-think-government-should-do-more-on-climate/

13 Braga Meininger, Blog, "The Years Beyond Youth" (May 21, 2023) at: franm.mgc@blog.wixnotifications.com

14 Earth Observatory, "Arrhenius" (Jan. 11, 2020) at: https://earthobservatory.nasa.gov/features/Arrhenius

15 Michael Shellenberger and Ted Nordhaus, *Break Through: From the Death of Environmentalism to the Politics of Possibility* (Houghton Mifflin, 2007).

16 David Pogue, *How to Prepare for Climate Change: A Practical Guide to Surviving the Chaos* (Simon and Schuster, 2021).

17 Hope isn't the same as optimism. But it's different from defeatism ("It's hopeless"). For analysis read Rebecca Solnit: "We can't afford to be climate doomers" at: www.theguardian.com/commentisfree/2023/jul/26/we-cant-afford-to-be-climate-doomers The NotTooLate website invites newcomers to the climate movement as well as people who are already engaged but weary, urging people to see why it's worth doing the work the climate crisis demands of us. See: www.nottoolateclimate.com/

18 Katharine Hayhoe is both a climate scientist and an Evangelical Christian, as she explains in her book *Saving Us: A Climate Scientist's Case for Hope and Healing in a Divided.* For my own treatment of religious responses to climate, see H.R. Moody and W. Andrew Achenbaum, "Solidarity, Sustainability, Stewardship: Ethics Across Generations" *Interpretation: A Journal of Bible and Theology* Apr. 2014). See also: Tim Alberta, *The Kingdom, the Power, and the Glory: American Evangelicals in an Age of Extremism* (Harper, 2023); and Mitch Hescox, Paul Douglas and Paul Douglas, *Caring for Creation: The Evangelical's Guide to Climate Change and a Healthy Environment* (Bethany House, 2016).

19 Janet L. Gamble, Bradford J. Hurley, Peter A. Schultz, Wendy S. Jaglom, Nisha Krishnan and Melinda Harris, "Climate Change and Older Americans: State of the Science" *National Institute of Environmental Health Sciences* (2012) at: http://ehp.niehs.nih.gov/1205223/

20 S. Cutter and C. Emrich, "Moral Hazard, Social Catastrophe: The Changing Face of Vulnerability along the Hurricane Coasts" *The ANNALS of the American Academy of Political and Social Science* 2006;604(1):102–112.

21 Michael Greenberg, *Protecting Seniors Against Environmental Disasters: From Hazards and Vulnerability to Prevention and Resilience* (Routledge, 2014).

22 John Kotre, *Outliving the Self: How We Live on in Future Generations* (W.W. Norton & Company, 1996).

23 James Hansen, *Storms of My Grandchildren: The Truth About the Coming Climate Catastrophe and Our Last Chance to Save Humanity* (Bloomsbury, 2010). See also: Larry Rasmussen, *The Planet You Inherit: Letters to My Grandchildren When Uncertainty's a Sure Thing* (Broadleaf Books, 2022).

24 "The Over-60 Crowd Steps Up with Bill McKibben's Climate Action Group Third Act" at: www.sevendaysvt.com/vermont/the-over-60-crowd-steps-up-with-bill-mckibbens-climate-action-group-third-act/Content?oid=37895972

25 Breanna Draxler, "The Power of Inclusive, Intergenerational Climate Activism" *Yes! Magazine* (Sept. 2020) at: www.yesmagazine.org/environment/2020/09/21/intergenerational-climate-activism/

26 "Former Head of NASA's Climate Group Issues Dire Warning on Warming" *Inside Climate News* (Nov. 3, 2023) at: https://arstechnica.com/science/2023/11/former-head-of-nasas-climate-group-issues-dire-warning-on-warming/

27 Tracey Skillington, *Climate Change and Intergenerational Justice* (Routledge, 2020).

28 G. Haq, J. Whitelegg and M. Kohler, "Growing Old in a Changing Climate. Meeting the Challenges of an Ageing Population and Climate Change. Stockholm" *Sweden: Stockholm Environment Institute* (2008) at: http://seiinternational.org/mediamanager/documents/Publications/Future/climate_change_growing_old.pdf

29 Per Espen Stoknes, *What We Think About When We Try Not to Think About Global Warming: Toward a New Psychology of Climate Action* (Chelsea Green, 2015).

30 Robert H. Frank, "Shattering Myths to Help the Climate" *New York Times* (Aug. 2, 2014).

31 Naomi Oreskes and Eric Conway, *Merchants of Doubt: How a Handful of Scientists Obscured the Truth on Issues from Tobacco Smoke to Global Warming* (Bloomsbury, 2011).

32 Stuart Bryce Capsticka and Frank Pidgeona, "What *Is* Climate Change Scepticism? Examination of the Concept Using a Mixed Methods Study of the UK Public" *Global Environmental Change* Jan. 2014;24:389–340.

33 Paul Hawken, *Blessed Unrest: How the Largest Social Movement in History Is Restoring Grace, Justice, and Beauty to the World* (Penguin, 2008).

34 Robert Butler, *The Longevity Revolution: The Benefits and Challenges of Living a Long Life* (Public Affairs, 2008).

35 Wilford Welch, *In Our Hands: Handbook for Intergenerational Actions to Solve the Climate Crisis* (In Our Hands, 2017).

36 Jonas Salk, "Salk Together: Celebrating a Legacy of Discovery" *Salk Institute* (1994) at: www.salk.edu/engage/your-impact/improving-lives/

37 Roman Krznaric, *The Good Ancestor: A Radical Prescription for Long-term Thinking* (The Experiment, 2020).

38 Ashton Applewhite, "How Did Old People Become Political Enemies of the Young?" *The Guardian* (Dec. 23, 2018).

39 For We Don't Have Time, see www.wedonthavetime.org/our-community For It's Not Too Late see www.nottoolateclimate.com/

40 Language selected and adapted from Robert Reich's newsletter "Welcome to (Gulp) 2024" at: https://robertreich.substack.com/Reich writes about Trump and I've focused on climate change.

Part IV

Actions

8 Becoming a Climate-Conscious Consumer

In the struggle over climate change individual consumers are important, and becoming a climate-conscious consumer is a task that falls upon all of us, young and old alike. New ways of measuring our climate footprint point to the role of consumption, which has a greater role in global emissions than we usually think.[1]

Consumer behavior is so big that it must be at the center of any discussion about how to hold back climate change and its potential to harm us. What some have termed the "Great Acceleration" of recent decades continues to gain speed, and its impact on our lives is not good. The result is what can well be described as a collision course of endless growth on a finite planet.[2] About a quarter of greenhouse emissions come from things that we buy, the consumer economy.[3]

The Question for Each of Us Is: Why Do We Keep Buying New Stuff?[4]

The bottom line is that if we want to "Make America Great" we have to begin at home, in the USA:

> The United Nations panel that studies climate change found that consumption surpassed population growth as the biggest driver of resource and material use on the planet at the turn of the 21st century. In the United States the population has increased 60% since the 1970s while the consumption of consumer items has increased 400%. We consume 13 times more than the average person in one of the poorer countries in the world. This means that having two children in the US is like having 20 children in the poorer countries of the world.[5]

The key argument of this book is summed up in four words: HERE NOW YOU HOPE. The point is nowhere clearer than when it comes to what we buy and consume, when it comes to how we live our lives:

> What kinds of things am I doing or buying without thinking about where they come from or what kind of impact they have? How have our habits and our expectations changed over time, maybe generating more waste or encouraging more consumption? What do I notice

DOI: 10.4324/9781003345992-12

myself and other people doing that seems wasteful but appears to happen without a second thought?[6]

To ask these questions is to begin to become a climate-conscious consumer. For those of us who are old, it comes down to what I, along with others, have termed "conscious aging."[7] It is, in short, a step toward freedom as well as a step toward caring for the planet around us. We are living in a hyper-consumer society, and the results for planet Earth are not good: "a loss of biodiversity and native ecosystems, poor soil health . . . more fertilizer (because of that poor soil health), which pollutes ground water, drinking water, oceans, rivers, and lakes."

Becoming a climate-conscious consumer is essential for confronting global warming and climate change. With shrinking polar icecaps and declining biodiversity, our shopping habits are one way that we as individuals can make a difference. We can hope for new technologies or renewable energy sources in the future. But how we use our credit cards is the choice available to us right now. In later life and even over the whole life course the new economics of true wealth can be very different from what the consumer economy taught us to believe.[8]

When it comes to aging, there are advantages to having bought all the things you need earlier in life. The aging consumer is a subject deserving more attention in its own right.[9] But people of all ages need to wake up to the importance of the consumer economy in an era of climate change: "Greenhouse gas emissions are the direct physical drivers of climate change [but] those gases do not just miraculously appear. Rather, they are the result of production and consumption systems."[10] Young or old, we are faced with the challenging question: Could there could be a reasonable hope that shifting consumer values can combine with market forces to green the economy?[11]

But the situation is not hopeless at all. It turns out there is a vast secondary market for used products of all kinds. It's available through the online "recommerce" ecosystem, and it is clearly a path to becoming a climate-conscious consumer. This ecosystem offers returned and refurbished items, and it includes providers like eBay, Etsy, Goodwill, and Amazon. Clothing, furniture, and electronics are in the picture. For those reluctant to buy entirely online, there are other channels, such as OfferUp, Facebook Marketplace, Nextdoor, and Craigslist.[12]

It is impossible in a limited space to cover all aspects of becoming a climate-conscious consumer. In this chapter we look at three major categories: home energy use, diet, and local transportation. The task is to become more conscious of what we do in all these domains, whatever our age.

One of the great legislative victories on climate change came in the summer of 2022 with the Inflation Reduction Act (IRA). About the IRA, we can say what Winston Churchill said after the second battle for El Alamein in Egypt (November 1942): "This is not the end. It is not even the beginning of the end. But it is, perhaps, the end of the beginning."

Churchill was on target in 1942 and, 80 years later, 2022 also marked the end of the beginning. As with WWII, the struggle will continue for years. Activists from the political movement known as Indivisible have said that protest work is not a sprint but a marathon. We have no time for anger or despair. It is time for slogging, or, as Peter Drucker once said: "Every great idea eventually degenerates into work."

Becoming a climate-conscious consumer will be at the heart of climate advocacy in the coming years, and people of all ages will be part of it. But the battle will take place not in central places but in a very dispersed way, in homes all around the USA. Bill McKibben[13] (2022) puts the challenge well:

> There are about a hundred and forty million homes in the United States. Two-thirds, or about eighty-five million, of them are detached single-family houses; the rest are apartment units or trailer homes. That's what American prosperity looks like: since the end of the Second World War, our extraordinary wealth has been devoted, above all, to the project of building bigger houses farther apart from one another. The great majority of them are heated with natural gas or oil, and parked in their garages and driveways or on nearby streets are some two hundred and ninety million vehicles, an estimated ninety-nine per cent of which . . . run on gasoline. It took centuries to build all those homes from wood and brick and steel and concrete, but, if we're to seriously address the climate crisis, we have only a few years to remake them.

McKibben points to one area in particular: home energy use, which means replacing furnaces and gas burners with heat pumps and induction cooktops. He also points to those cars in the driveway, the primary form of local transportation. What will it take to move to more electric vehicles (EVs)? We can add to this list the fact that eating is also a big part of consumer behavior contributing to greenhouse gas emissions. What will it take to change our diet in the struggle around climate?

A key point here is that government policy and the private marketplace alone will not guarantee a change for decarbonization, not unless consumer behavior responds to what is needed for a sustainable climate. For older people, this change may entail changing a lifetime of habits. As gerontologist Robert Kastenbaum suggested, the best definition of aging is habituation—the habits that we adopt and don't change. But for people of any age, habits can be changed. In this chapter, we look at becoming a climate-conscious consumer in an aging society.

Can one person make a difference? Yes, but it depends on what we do, and the answer isn't always obvious. Making a difference requires some inquiry, and that's what becoming a climate-conscious consumer is all about. The biggest barrier is internal, in our own head. Changing one's own mind is not only possible; it's indispensable. When it comes to climate change, the message is clear: it's not too late and you're not too old. Here are some steps that are within our power if we want to make a difference:

Eliminate Food Waste

Food production systems of all kinds contribute perhaps a third of the green-house gases driving climate change. The National Defense Resource Council estimates that Americans throw away up to 40% of food purchased.[14] There are lots of ways to eliminate food waste, such as buying only what you need, "recycling" leftovers, composting food scrap, and donating what you can to food banks.

1. **Eat Plant-Based Food.** You can aim for a vegetarian or vegan diet, but that step isn't necessary to begin to make a difference. An Oxford study suggested that going to low consumption of meat will reduce the foodprint by up to a third. Reducing meat consumption and deforestation on a global scale would have an enormous impact. There are many simple things we can do, at any age. Sustainable diets are familiar, but there are new approaches, such as a cookbook giving guidance on food and climate change.[15]

2. **Use Green Energy.** Renewable energy is key to a green world. Putting solar panels on the roof is great, but people in condos (like me) or renters may not be able to do that. It's often possible to buy renewable electricity from other sources. Options for green energy exist if we start looking for them.

3. **Better Insulation.** Energy costs can be reduced beyond the purchase of green energy. Start with insulation. Older homes can lose more than a third of their heat through the walls. As with avoiding food waste, it's possible to respond to the climate crisis by better insulation, reducing greenhouse emissions, and saving money at the same time.

4. **LED Lighting.** Another place to reduce energy waste is in the lightbulbs. LED bulbs use up to 90% less energy than the incandescent bulbs we grew up with. LED bulbs also mean lower costs for air conditioning because they produce less heat themselves. The bulbs also last longer, offering another path to saving money.

5. **Local Transportation.** We know that shipping and airlines contribute massively to greenhouse gas emissions. But local transportation is much more within our power. We can support public transit, where possible, and, when buying a new car, we can choose an electric vehicle, which saves money and reduces greenhouse gas emissions. Gas-powered vehicles of all kinds account for 7% of global greenhouse gas emissions. For that reason, transitioning to EVs could be a major help in reducing greenhouse gas emissions.[16] The gas-powered cars we elders grew up belongs to the past, not to the future.

6. **Recycle.** Recycling doesn't solve all problems, but it can make a difference. Around half of all recycled materials come from households, and increasing that proportion would make a difference. For example, food waste in landfills creates methane, a greenhouse gas more damaging than carbon

dioxide. Recycling successfully depends on learning how to do it correctly. But doing it right can make a difference.[17]

7. **Buy Less.** We've heard about the new three R's: Reduce, Reuse, Recycle. Some would now add a fourth R: that is, refuse to buy what you really don't need. As people get older, they typically find they don't need to buy certain items, like big household appliances. But, along with discretionary time and income, older consumers may have the option to buy more than they need or want. In our consumer society, it takes commitment and knowledge to take a different approach. A tool such as footprintcalculator. org can help measure one's ecological footprint.

These eight steps are drawn from Green America's action steps, documented in the work of Project Drawdown.[18] Green America also offers guidance on purchasing products and services by becoming a climate-conscious consumer.[19] Other steps, such as divestment from companies or participation in the democratic process, are discussed in subsequent chapters of this book.

But the message here is that climate advocacy begins at home. Becoming a climate-conscious consumer has financial benefits for consumers. The reason is simple: many items of home consumption have increased in cost because of climate change. Examples here include grocery bills, where prices have been driven up by so-called heatflation. Other examples include water bills (remember drought) and electricity costs.[20] But it turns out that, with small efforts, it's possible to dramatically reduce the amount of plastic packages and instead go for glass or cardboard. Small steps are possible.

Those who are old know their time on the earth is limited, so they may think: What's the point? But how many people go around thinking "I'll be dead in five years." If you think that way, it would be wise to see a psychotherapist or a hospice counselor, depending on the situation. True, five years is in the future, but it isn't a long, long time. It's time enough to get payback for lots of consumer choices to save money under the IRA. In this chapter, we focus on just a few of the big categories: nutrition, home energy options, and local transportation.

Becoming a climate-conscious consumer starts at home: with the things you buy, the food you eat, and all the purchases of goods and services that make up your personal economy. The problem is that, as consumers, we consume a lot and we consume many different things. Consumption changes over the life course: for example, older people who are aging in place are less likely than young people to buy new appliances, so energy-saving questions may be of less interest to the old than to the young. But there's a bigger problem. Research has found that even people who want to be client-conscious consumers may have mistaken ideas about the impact of buying on climate change.

Researchers have found that people often adopt habits that may seem significant but actually have a very small effect on limiting climate change.

A valuable article on this subject is titled "What's the Best Way to Shrink Your Carbon Footprint?" Even the term "carbon footprint" is an irony: the phrase itself was invented by an oil company, British Petroleum. The phrase is reminiscent of other marketing efforts by petrochemical companies. Devising a phrase like "carbon footprint" was part of a strategy to suggest that individual actions—total carbon footprint—are the big question to focus on, rather than structural questions like transportation policy or tax policy.

The key point is that people have very little idea about the impact of individual consumer decisions on limiting climate change. For example, recycling, carpooling, or lowering the room temperature has only a small impact, while living car-free or eating a vegan diet has a very big impact. Recycling or changing the thermostat is easy enough to do. Doing without a car or trying to eat a vegan diet is hard. We need to take a more incremental approach: How to "nudge" people to think better about climate change choices?[21]

In this chapter we will look at only a few consumer choices where becoming a climate-conscious consumer has a significant impact and is also feasible to do. We will look specifically at diet, using an electric car, and energy use for heating and cooling at home. What actions are needed? What actions are possible? As a rule, researchers have found that Americans underestimate the impact of actions that are hard to undertake—like eliminating air travel or radically changing their diet. They tend to overestimate the impact of actions that are easy to undertake. But it remains true that it's better to light a candle than to curse the darkness. On climate change, we begin where we are and we do what we can.

Is Consumer Action the Solution?

No, it is not the whole solution but it is part of the solution. Here lies the problem with talking about climate change in terms of consumer action and becoming a climate-conscious consumer. The problem is that it keeps the discussion at the level of individual choices. There's a long history of this kind of framing in the environmental advocacy movement. In 1971, for example, not long after the first Earth Day, we saw lots of ads on TV and in newspapers, ads paid for by big corporations, urging us to work against pollution. One of the most prominent had the image of a fiercely staring Native American man, with the message: "Pollution: It's a Crying Shame. People start pollution. People can stop it." In this framing of the problem, people—that is, individuals—are the beginning and end of the problem (start it–stop it). There is no place here for companies that produce products that end up as pollution or end up warming the earth's atmosphere. There is no place here for the government to take action that could prompt businesses and individuals to make choices that will address the problem.

Where to begin? Start with three areas where you act as an individual: as a consumer (buying things in the marketplace); as an investor (holding financial assets for the future); and as a citizen (acting to influence government).

Figure 8.1 "Get Involved Now. Pollution Hurts All of Us." Keep America Beautiful Inc. Public Domain.

We are, each of us, consumers, investors, and citizens (whatever words we use to describe these discrete roles).

There are good and bad ways of being in these roles. "Good" for being a consumer means getting what you want, getting it in the most efficient (economical) way, and getting what's good for you (whether you want it or not). In diet, "good" means avoiding things that are bad for your health and consuming foods that are good for your health. People often don't have the means, economic, intellectual, logistical, geographical, and so on, to make choices even if they know they should or want to. This book is intended to clarify those real-world choices for consumers in three areas: home energy use; nutrition; and local transportation. There are other areas (e.g., air transport) left out of this chapter because these are not as amenable to individual choice or action as the ones considered here. David Wallace-Wells put it directly in *The Uninhabitable Earth: Life After Warming*: "The climate calculus is such that individual lifestyle choices do not add up to much, unless they are scaled by politics."

What can you as an individual do about climate change? A survey found that nearly 60% of people answered "recycling." But it turns out that recycling isn't a particularly significant way to impact climate change. Imperial College London listed the top nine things you can do (Grantham, 2023). One of them is to invest your money responsibly, something discussed in the chapter on investment for a green retirement. But others are consumer actions:

Eat less meat and dairy
Reduce energy use, and bills
Cut consumption—and waste
Reduce driving

Still others are discussed in the chapter on citizen action on climate policy, and that means making your voice heard by those in power and talking about the changes you make.[22]

Becoming a climate-conscious consumer is easier than think—but it's also harder than you might assume. It's easier because you can do it every day—with diet, for example—and also with more expensive actions, like home energy enhancement, which saves money. But it's also harder than one might expect because when we enter the marketplace—any marketplace—we run into the problem of getting accurate and unbiased guidance. In the following chapter, on retirement investment, we see how this challenge unfolds in the financial industry.

For consumer purchases, think only of the example of buying "organic" foods. More than 30 years ago the term "organic" was a new idea. By 1990 some farmers joined forces to oppose industrial farming practices that gave us pesticides in the food supply and antibiotics in livestock. Problems still exist, but the U.S. Department of Agriculture at least has standards for what can and cannot be labeled "organic." Companies can still claim foods are "All-Natural," and some buyers will be swayed by that claim. But those looking for genuine organic food can find the help that they need. We will all need to apply this kind of critical intelligence in the marketplace to become a climate-conscious consumer.

Could we do the same for consumer products where individual choices could make a difference around climate change? The answer is yes: it's easier than you think and it does make a difference:

> In fact, when added up, thousands of small actions, like switching off lights during a heat wave, buying efficient appliances, or voting for politicians who will act on climate change, can become a larger force than what governments and businesses can muster on their own. Put another way, a comprehensive approach to climate change demands actions, both large and small. "It's a mistake to only focus on governments and big companies," said Paul Burger, head of the sustainability research group at the University of Basel, who studies consumer decisions on energy. "It's also a mistake to only focus on individuals."[23]

Home Energy

How can we get to a sustainable future on the earth? The answer lies in 100% clean electricity or as close to that goal as possible. But getting there soon will not be easy. First, we need clean electricity from clean power sources. Second, we need to transmit the power to end users. These end users include machines, buildings, and industries but also consumers in their own homes. Combining both production and consumption of energy could cut up to 75% of greenhouse gases in America. That means that electrification is the best opportunity between now and 2030 for reaching the U.S. pledge of a 50% reduction in greenhouse gas pollution. Individuals, of whatever age, have limited ability to control electricity production or transmission. But we have great ability to control energy in our homes and that is the focus of this discussion.[24]

The issue of total energy use is particularly important for Americans, as Figure 8.2 makes clear:

The Rocky Mountain Institute maintains that energy overall is 70% of the problem of climate change. Seventy-six percent of global greenhouse gas emissions come from energy (electricity, fuels, heating, etc.), mostly produced by fossil fuel companies and backed by governments.[25] Indeed, around 84% of our energy comes from the burning of fossil fuels. And there's the problem: The global energy sector remains 80% dependent on fossil fuels. An important path to a solution will be the electrification of home energy.[26]

Heating and cooling the home is one of the biggest points where one can become a climate-conscious consumer: "Heating and cooling are the largest energy expenses in most homes–comprising 35–50 percent of the annual electricity bill. They also account for a significant portion of an individual's impact on the climate."[27]

As with many consumer decisions, doing the right thing means becoming well informed and often navigating a confusing range of choices and constraints: for example, local building regulations. In any case, before replacing a fossil-fuel-fired furnace, it would be wise to weatherize the home to prevent the loss of energy because of poor insulation and design. There are businesses emerging around the country that can help guide consumers through the challenging process of home electrification. But, again, choosing the right contractor is not simple.[28]

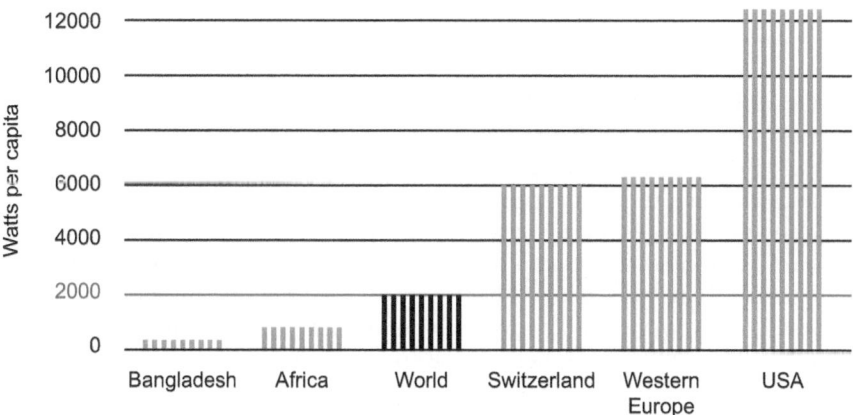

Figure 8.2 Energy usage in watts per capita. Adapted from a graphic by The 2,000 Watts Society.

A leading group pushing for home electrification is Rewiring America, founded in 2020. The group promotes replacing gas heaters with reverse-cycle air conditioners, induction cooktops, heat pumps, and other changes in home energy consumption. Widespread adoption of highly efficient heat pumps could make a big difference.[29]

There are major advantages to the IRA, and Rewiring America has developed a calculator to show the multiple tax credits and rebates now available.[30] Norway has installed heat pumps in two-thirds of households in the country, showing that switching to greener heating can be done.[31] Can't afford to replace your home energy with a heat pump? It could be easy enough to reprogram your thermostat.[32]

Saul Griffith, founder of Rewiring America, says, "40 to 42% of all of our emissions in the domestic economy come from decisions that are made around the kitchen table." Griffith's vision is squarely in the framework of becoming a climate-conscious consumer, starting with consumer demand:

> You never hear that talked about in traditional climate action. You hear about boycotting oil or shutting down the coal station. But we have to decarbonise this demand side at the same rate that we decarbonize the supply side, or it doesn't add up.[33]

Inflation Reduction Act

For those who want to take advantage of tax credits and rebates under the IRA, the first step should be to carry out a home energy audit. There are many home energy audit companies that can provide such services, but the challenge is to find one who is local and who is reliable, without any conflict of interest such as trying to sell products or services.

How Much Could One Save If an Energy Audit Points to Changes in the Home?

> Overall, you would qualify for a 30% tax credit from the Residential Clean Energy Tax Credit. They might suggest you install solar panels on your roof which would mean up to $9,000 saved on installation, and less time to see a return on investment as well. Your home audit also might suggest that you buy new, energy-efficient appliances, which have an annual $2000 cap through the RCE Tax Credit. These appliances include heat pumps, heat pump water heaters, biomass stoves and boilers, with a $600 cover cap per item. One stipulation is that the renovations must have been made to a non-commercial home, which also covers renters and multifamily buildings.[34]

By some estimates, an average household could expect a net benefit under the IRA of $1,000.

This analysis is compelling, but it doesn't note where some of the biggest savings are possible: namely, in eliminating waste by improving insulation and avoiding leaks. Once this is done, it makes sense to consider more extensive electrification such as improved electric wiring or solarizing a home through rooftop solar panels. Upgrading homes for energy efficiency may also include rebates available under the High-Efficiency Electric Home Rebate Act, which could offer up to $14,000 discount for home electrification. But, again, the details to qualify for rebates and the time period for payback are critical factors to consider.

Becoming a climate-conscious consumer depends in big part on the IRA:

> For all that has been written about the Inflation Reduction Act, the most salient fact about it remains widely underappreciated. What is significant about the bill is not just that it sends an enormous amount of money toward climate solutions, but that the money is almost entirely uncapped.[35]

The point here is that the amount of government money in the IRA isn't specifically limited but depends on the demand for tax credits. The more people who ask for them, the more gets spent. What are the numbers here? The Congressional Budget Office estimates that the IRA will spend $391 billion, but a Goldman Sachs report put the number as close to $1.2 trillion. New York Hotel tycoon Leona Helmsley once said that only a little people pay taxes. If the "little people" apply for tax credits under the IRA, it will have a big impact. But getting people to understand how to become a climate-conscious consumer, and get a payoff, will depend on educating and mobilizing people.

Electrification

There are many benefits to electrification, including health, safety, comfort, and resilience. Residential buildings account for a large percentage of greenhouse gas emissions. The net cost of replacing an obsolete gas-fired stove, water heater, dryer, and furnace can be substantial: for example, $24,000 compared to $16,000 for replacing these units with gas. Wait: isn't $8,000 a lot to pay to go all-electric to reduce climate change?

But the picture changes when you include rebates and the IRA of 2022, which can change the game completely. In the case just given, the cost would drop from $24,000 to only $10,000 when you include rebates and tax credits. That figure is $6,000 less than the $16,000 figure that it would have cost to get gas replacements. In other words, electric replacement can be less than gas replacement.[36]

A key point here is that the benefits of so-called natural gas (methane) in a power outage are exaggerated by the fossil fuel industry, along with exaggerated fears about electrification. Gas central furnaces, newer stoves, dryers, and gas water heaters will not operate during power outages. It's not necessary for older homeowners to tear out appliances immediately. All gas-fired equipment will eventually fail and need replacement. Why think of replacing gas equipment with gas, when many clean electricity options are available at competitive costs? Working against climate change can be a money-saving proposition. And the path begins at home.

Gas Stoves

Methane, generally called "natural gas," governs much of what happens in our homes, from cooking to staying warm. We depend on gas, but it often makes our lives worse in ways we cannot even see. Methane produces damaging pollutants that can lead to multiple disorders. It is estimated that gas consumption contributes the majority of greenhouse gas emissions.[37] Becoming a climate-conscious consumer begins at home, and the first issue we address here is home energy use.

For climate-conscious consumers, one of the biggest places for home energy attention is the kitchen, above all the stove that cooks our food. Leading consumer safety officials have called gas stoves a "hidden hazard." They have reasons to say so even apart from impact on the atmospheric environment. There are damaging health consequences for consumers from nitrogen dioxide and particulate matter created from gas stoves. The 2022 IRA provides money to replace these stoves with electric ranges or induction cooktops. Rewiring America provides a valuable calculator to see more about these options and learn what savings are possible.

Testing of induction ranges finds superior performance along with a lower likelihood of health risks. In testing by Consumer Reports, the group found that electric and induction options were at the top level of efficiency. Consumer Reports found that induction ranges and cooktops heat fastest and deliver rapid temperature changes when the burner is adjusted. In the Consumer Reports rating, 80% of induction ranges performed well enough to be recommended, in contrast to fewer than half of the gas ranges tested.[38]

American homes are about equally divided between electric and gas stoves. Electric stoves avoid using fossil or natural gas and also avoid releasing pollutants into indoor air. Induction cooking is a further refinement of electric cooking because it uses direct induction to heat the cooking vessel itself. Induction stoves are more efficient than ordinary electric stoves but significantly more efficient than gas stoves. One problem is that induction cooktops cost more (e.g., $1,000). There is public interest but only 3% of households have induction stoves.[39]

Many people do consider moving away from gas stoves to electric or induction stoves, and they have good reasons.[40] Gas stoves emit nitrogen oxides and other pollutants into indoor air. Such indoor pollution can contribute to asthma in children and respiratory problems in older people. Should older homeowners replace their gas stoves with an induction stove?

The answer isn't simple.[41] Induction stoves offer major benefits over gas, but home cooks face barriers in making the switch. Switching from gas to electric stoves may be good for climate, but, predictably, the issue has also inspired a conservative backlash, with valid concerns. For example, older buildings can pose a big problem for upgrading to electric appliances. As with giving up a gas-powered car for an EV, older people have "sunk costs" if they have a perfectly functional gas stove and they may be reluctant or unable to generate the substantial upfront cost for a new stove, unless they have to replace the stove anyway. There is also the lifelong preference for gas cooking and the "love of fire" as a common human phenomenon.

Gas stoves emit methane, a potent greenhouse gas: up to 80% of climate-damaging methane comes from gas stoves even when they are not turned on.[42] According to the International Energy Agency, methane is responsible for around 30% of the current rise in global temperature.[43] To that in perspective, methane leaks from American gas stoves, mostly when the stoves are off, in a year have the same warming impact as carbon dioxide emissions from 500,000 cars.[44]

In sum, there are both environmental and personal health reasons for a climate-conscious consumer to consider moving to an electric or induction cooking system. The public debate will not soon disappear and personal consumer choice will play a part here.

Many consumer issues can easily get politicized and become part of culture wars.[45] For example, the Green New Deal pointed to the role of greenhouse emissions on the farm, such as beef from cattle. The idea soon became framed as "They're taking away your hamburgers!" Older consumers may be reluctant to move to induction cooking because of upfront or other barriers. But for those able to make the move, there are clear advantages in terms of cost and health.

Is it worth it to get an induction cooktop stove? It could be three times more efficient than a gas stove, but only 10% more efficient than a conventional electric stove. In addition, the cost of installing an induction stove isn't cheap: $700 or up to $,2,000. It's also likely that new cookware will be needed because aluminum, glass, pyrex, or copper doesn't respond to magnetic forces. The cost could be a deal-breaker for many people. Consumers need guidance on how to find the right model of induction cooktop to buy.[46] Older people on a fixed income may feel they can't afford to replace their home energy with a heat pump. But it's easy enough to reprogram your thermostat, which could make a big difference.[47] As with IRA incentives, there are opportunities for doing well by doing good. When it comes to becoming a climate-conscious consumer, the Buddhist teacher Pema Chodron said best: Begin where you are.[48]

DIET FOR A SMALL PLANET[49]

Health, diet, and climate are connected. Start with health and the bad news: two-thirds of American adults are overweight, and over a third are obese.

Others suffer from chronic conditions, such as high blood pressure, that are exacerbated by diet. The reasons aren't hard to see. Grocery stores contain salty crisps, sugary drinks, and processed foods of every variety. The problem has gotten worse in recent decades, exactly the same time when climate change has gotten worse. No, climate change alone didn't cause bad health. But each year, as we get older, the impact of bad food gets worse and agricultural practices damage the environment and intensify climate change. Food consumption is part of the picture. Becoming a climate-conscious consumer will have health benefits as well as a positive impact on climate. As with climate, so with diet: You are not too old and it's too late.

What we eat has a big impact on climate change,[50] and it's a key part of becoming a climate-conscious consumer.

There are some general trends to note. For example, sugary beverages, fruit, and bread tend to have a lower impact on the environment, while meat, fish, and cheese have a greater impact. Deserts and pastries are somewhere in the middle. The good news is that, while individual products differ, in general, more nutritious foods also tend to be more environmentally sustainable. By some estimates, food accounts for more than a third of total greenhouse gas emissions.[51]

Some things about food consumption are simple: for example, avoiding waste. Scrapping use-by dates could prevent huge amounts of food waste.[52] But it requires some learning to assimilate the ABC's of managing food waste.[53]

A key principle for a sustainable diet is to shorten as much as possible the number of food miles—the distance foods have traveled to get to your home. Eating locally is a good idea. Farmer's markets, where available, are one practical approach here. It's also vital to understand who produced the food and what is their credibility. On this point the Sustainable Food Trust is a most helpful source for guidance.[54]

> Good eating starts at home, and one of the most important things we can do for the future of the planet is to minimize food miles—so our staples should be foods that can grow perfectly well in this country," advises Patrick Holden, chief executive of the Sustainable Food Trust. Another basic principle is to do your best to understand the story behind what you're eating—be it plant or animal: "If you know who produced your food, they are accountable to you, and more likely to care.

Diet and food choices are in control, aren't they? The answer is yes and no.[55] We do have power of agency, but there are limits. When it comes to food, becoming a climate-conscious consumer is not as difficult as people think.[56]

The whole framing of what constitutes "good" and "bad" food is shaped by cultural and community meanings. We can't talk about individual choices and responsibilities for health outcomes without taking into account economic and social inequality. Do people who are poor have a real choice about buying "organic" or "regeneratively grown" food? Leah Thomas put it this way: "We can't save the planet without uplifting the voices of its people, especially those most often unheard."[57]

Besides income, *where* people live matters, as in the saying "Zip Code is destiny." Living in a so-called food desert means that individual choices matter less. It's possible to know what "good food" is, but the food we want turns out to be laden with environmental and manufacturing contaminants. Becoming a climate-conscious consumer, when it comes to food, is not as easy as we would like it to be.

What's good for the environment can also be what's good for our individual health. For example, preventing diabetes with good nutrition sometimes lies within our power. Educating people about this should be a part of health education over the whole life course, from childhood into old age. As with climate change, it's never too late to make changes. We should be educated about how to reduce "bad" foods and about what you can do both for the climate and also for your blood sugar. Focusing on what individuals can do

The Colorado River is being drained to produce beef and dairy

79 percent* of the river's water goes to crop irrigation. Here's how much each crop receives:

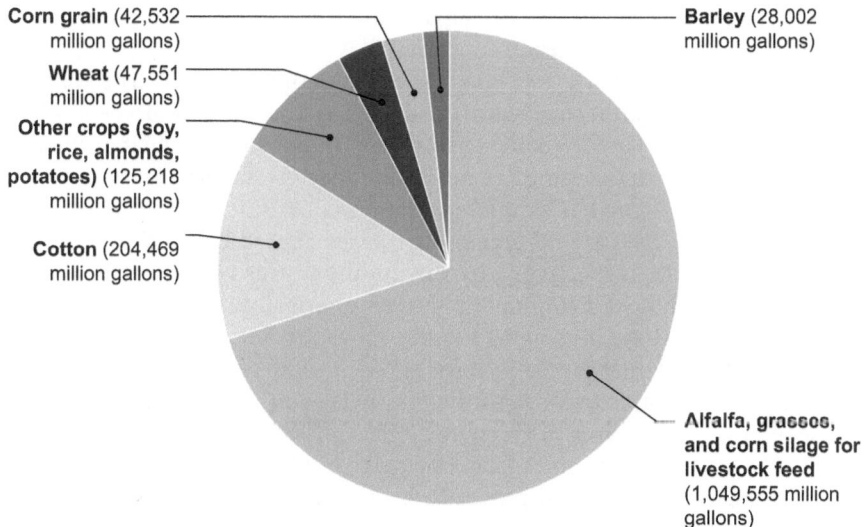

Corn grain (42,532 million gallons)
Wheat (47,551 million gallons)
Other crops (soy, rice, almonds, potatoes) (125,218 million gallons)
Cotton (204,469 million gallons)
Barley (28,002 million gallons)
Alfalfa, grasses, and corn silage for livestock feed (1,049,555 million gallons)

*Estimates for the Colorado River basin include only uses within the basin and do not include water exports from basin, reservoir evaporation, or natural losses.

Chart: Kenny Torrella/Vox • Source: Nature Sustainability

Vox

Figure 8.3 Usage of the Colorado River for agriculture. Adapted from a graphic by Kenny Torrella/Vox using data from Nature Sustainability.

is NOT at the expense of collective, public action. But, whether in climate action or better eating, the key is to overcome habits of mind that prevent action: for example, "It's too late" or "What can one person do?" When it comes to becoming a climate-conscious consumer of nutrition, the answer is still: HERE NOW YOU HOPE.

According to the United Nations, the world food supply is likely to be threatened if we don't take action to address climate change and its threat to food systems. What can one person do in response to the climate crisis? Well, the action begins at the dinner table. A long-standing body of research has shown that meat and dairy products are harming the environment.

Why is reducing meat consumption so important for climate change? The answer is that it's all about water. Let's begin with the biggest cause of the West's water crisis.[58] The Colorado River was discussed in the previous chapter on drought. But the biggest part of water, nearly 80%, from the Colorado River goes to crop irrigation. And of that water, the vast majority goes for alfalfa, grasses, and corn for livestock feed: in other words, meat and dairy.[59]

The alternative would be plant-based food, including fruits, vegetables, grains, and beans. It's important to recognize that this is not an all-or-nothing affair. We don't need to become "card-carrying vegetarians." Just reducing meat and dairy in favor of plant-based foods will have an impact. In other words, climate-friendly eating can make a difference, even if people don't go "all the way" on a vegetarian or vegan diet. You don't need to go vegan to significantly shrink your carbon footprint.[60]

The point is especially relevant for older people. It's not true that all older people are "set in their ways," but habits are valuable. Except when they're not, which is why reducing our habit of consuming meat and dairy is a habit worth challenging, at any age. The Physicians Committee for Responsible Medicine has estimated that a major shift to plant-based eating could reduce deaths and greenhouse gases caused by food production by 10% and 70%, respectively, by the middle of this century. The United Nations Environment Program puts it emphatically: "Animal products, both meat and dairy, in general require more resources and cause higher emissions than plant-based alternatives".[61]

It turns out that cows raised for milk, beef, and other products are the cause of nearly two-thirds of greenhouse gas emissions from the world's livestock. The largest share here is from one particularly damaging gas: methane, which animals burp out and proves to be a more damaging greenhouse gas than carbon dioxide.[62]

We could all eat less meat. It turns out that a meat-based diet demands 30% more water than a vegetarian diet. Close to 90% of Americans do eat meat, but meatless Mondays—that is, each person eating meat one less day a week—could save an amount of water equal to the entire annual flow of the Colorado River each year. That's one answer to the question, What can one person do?

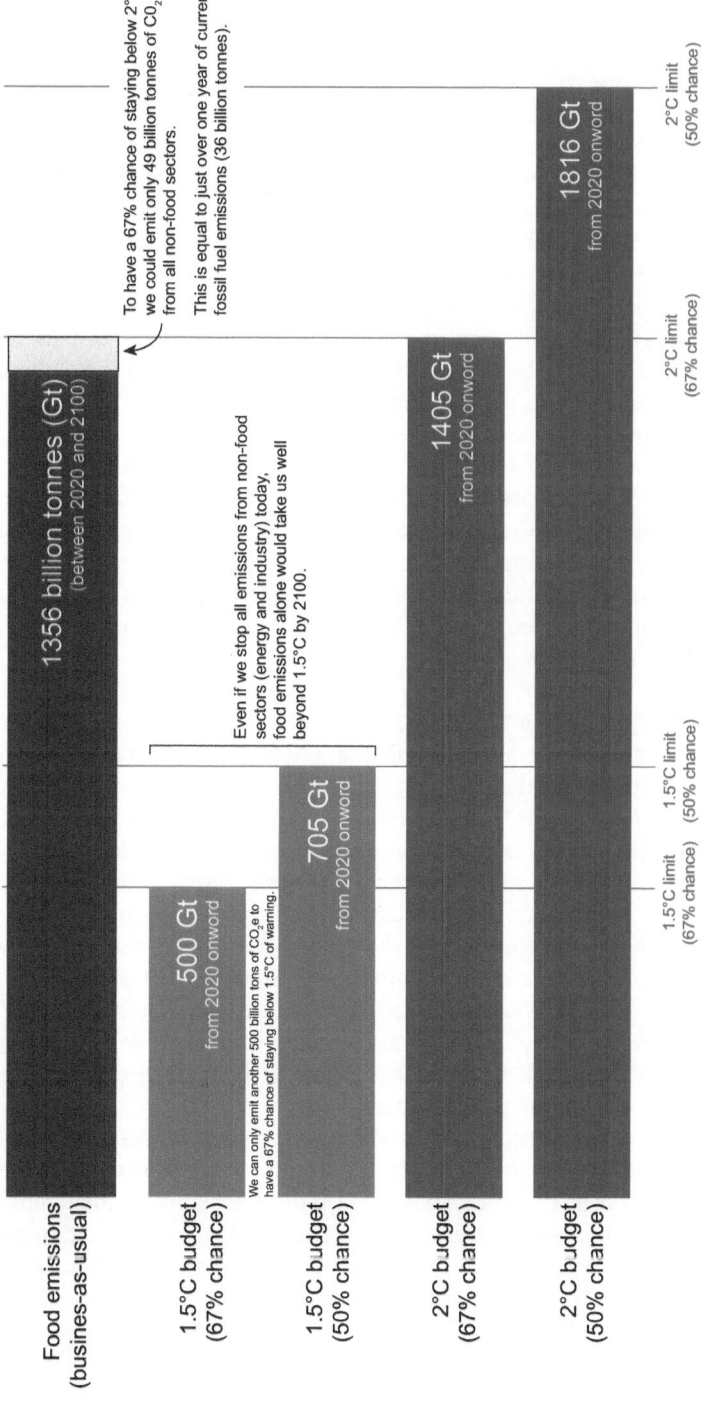

Figure 8.4 Impact of food emissions on the carbon budget. Adapted from Our World in Data.

We've gotten accustomed to knowing how the fossil fuel industry is subsidized, in various ways, by public policy and tax structures. The food industry, too, is a channel for self-destructive behavior, not recognized as such.[63] It may be time to get accustomed to the ways in which agriculture, and meat production in particular, is in a position where prices are kept low and public demand is kept higher than it would be otherwise.[64] Today's older population is directly implicated in this trend. Fifty years ago beef cost more than $7 a pound in today's dollars, while today, despite high inflation, the price of beef has come down to $4.80 per pound. Poultry is cheaper, but the key point is that low prices are hiding the genuine cost of meat.

Low food prices deceive us about the real cost of the food we eat: "The negative effects (pollution, emissions, etc.) of food production aren't reflected in the cost of food, but we are the ones who end up paying for them, in terms of health effects and climate change and pollution."[65]

Climate change is related here because animal agriculture likely contributes at least 14% of all greenhouse gas emissions and by some estimates much more. Meat production uses around half of all habitable land on the earth. The facts about climate change could encourage us to significantly reduce meat consumption, especially red meat with its problematic health impact. Becoming a climate-conscious consumer means thinking more critically about meat consumption. One example of this would be to favor meat produced through regenerative grazing.[66]

People who want to be climate-conscious consumers may imagine that they have to become vegans to make a difference. But going vegan isn't an option for everyone, and that includes older people who may have lots of reasons not to change their diet dramatically. We don't need more "judgemental veganism" and guilt.[67] Perfectionism and aiming too high in diet can be a mistake. It's not necessary to eat exotic foods like kelp or insects. Just eating less meat can make a big difference.[68] Recent research looked at a variety of "nudges" to encourage diet changes and investigators found that some simple and cheap ones can reduce the carbon footprint by up to 76%.[69]

The good news is that consumer choices about diet are already having an impact on climate. Over the past two decades, we have been seeing a gradual shift away from beef in the U.S. diet, and this trend will be reflected in the aging population. That trend has meant a 35% decrease in dietary carbon emissions during the past 15 years.[70]

For a climate-conscious consumer, what foods are best to eat? The answer isn't simple. We might just say cut down on meat or dairy and move toward a plant-based diet. But not all plant-based diets are equal in their climate impact, according to research conducted over a 30-year period.[71] It turns out that, in general, environmental impacts, including climate change, are correlated with health. What's healthier for you is better for the climate and brings fewer greenhouse gas emissions. But the relationship between food, health, and the environment is not simple. As with other questions for a climate-conscious consumer, getting the right answers will require study and attention.[72]

In this discussion we've been looking at climate-conscious consumer choices, while ignoring the wider political context. But if, for example, meat-eating has this big impact on climate, why not adopt policy choices that nudge consumers in a more climate-conscious direction? The recent IRA—the biggest federal climate legislation in history—largely ignored this dimension of diet and consumer choice Critics have argued that this historical legislation should be seen as a missed opportunity to cut food-related emissions. But critics recognize that the "politics of meat" is treacherous terrain. Opponents of action on climate have used provocative political tactics, such as claiming that advocates of climate action are "trying to take away your hamburgers."

> They found that products with a lower environmental impact tended to be more nutritious too. There were some exceptions: sugary beverages are one example of a product with a positive environmental ranking but negative nutritional value. But many other foods—fruits and vegetables, cereals, certain breads, meat alternatives like tofu and vegan sausages—were a win-win for both environmental and human health. This suggests "there does not need to be a tradeoff between nutrition and environment," the researchers write in their study.[73]

Food Waste

Thirty percent of all the food produced on the earth each year goes to waste, and that wasted food has an adverse impact on climate because of greenhouse gases.[74] If all the food wasted were considered as a separate country, the wasted food would rank as the third largest carbon dioxide producer in the world, bigger than the greenhouse gases produced by entire countries like India or Russia.

It is estimated that a quarter of household food is wasted. A third of that household food waste is caused by people cooking too much or serving too much. But two-thirds are due to food spoilage. When it comes to food waste, we're talking about a volume greater than plastic, paper, metal, glass, or rubber. Food waste is a big deal.[75]

Does food waste in America have to be this big? No, it doesn't have to be that way and we know this from comparing the USA with other advanced industrialized countries that have almost zero food waste.[76]

But the situation is far from hopeless. Increased consumer awareness has actually made the rate of food fall: between 2007 and 2012, household food waste went down by 15%.[77] This point is particularly important because so much of greenhouse gas from food waste comes from individual consumers at home:

Some food waste happens because food is transported for long distances: eating locally is a way to help reduce climate change. But a great deal of food waste happens at home, and it is within our power to change that.[78]

How do we avoid food waste? By changing habits in small ways. For instance, keep healthy snacks, drink more water, use whole grains instead of refined grains, and eat more vegetables and less processed meats. Above all,

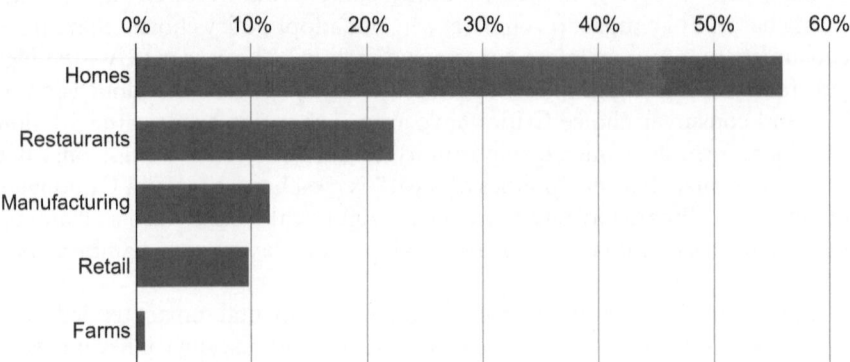

Figure 8.5 Greenhouse gas emissions from wasted food (2019). Adapted from Michael J. Coren/The Washington Post using REFED data.

make changes slowly, adopting food practices that will persist. Between 6% and 8% of all anthropogenic greenhouse gas emissions could be reduced if we were to stop wasting food.[79]

Another dimension of avoiding food waste is with composting. We don't have to feel obliged to eat every last scrap on the plate or in the food preparation process. Composting food scraps results in 38% to 84% lower greenhouse gas emissions than putting these food products into a landfill. Yes, trash goes to landfills, but heaps of compost can be watered and turned over, which puts air in contact with decomposing waste and thus stops bacteria from putting out methane, a greenhouse gas more potent than carbon dioxide.[80] I was keen on composting but downsized into multifamily housing and couldn't easily do it anymore, until the State of California put legislation in effect to make housing like mine put composting in place for everyone.

Sometimes small educational efforts make a difference. For example, only a minority, less than 40%, of consumers understand the difference between "sell by," "use by," and "best before" labels.[81]

Becoming a climate-conscious consumer begins with small steps. It doesn't end there. But our condition is a far cry from what "learned helplessness" about climate change might make us believe.[82] About climate change, and many other things, people often feel helpless and powerless. That feeling leads to complacency and paralysis: "Why do anything? It's all the fault of the rich, and I'm middle-class." But the middle class is important, too.[83] Consider this example. Americans who earn at least $40,000 a year might think of themselves as "middle-class." Well, think again. On a global scale, those earning more than 40 K are in the top 10%. More of us than we realize on a global scale are among the wealthy. More of us than people realize can make a difference when it comes to climate change.

We can start with small things. For example, because there are no national standards for food expiration dates, a lot of perfectly good food gets thrown away out because of misunderstood labels.[84] Avoiding food waste is a case where climate-conscious consumers can act, regardless of their income level.[85]

The bottom line, for both young and old: Smart Shop: don't buy more than you need; don't be deceived by expiration dates; and compose when you can. In short, there are many ways to reduce, reuse, and recycle food that otherwise would become waste.[86]

Reducing food waste could have a huge impact on global warming. Repeated studies have shown that a big proportion of food bought by Americans is simply thrown away. But reducing food waste by half is entirely possible. The average American family loses $1,600 every year—just by discarding usable foods.[87]

One of the easiest and most effective things climate-conscious consumers do is to reduce food waste. There are many things that individuals can do to keep wasted food, which is largely edible, out of the trash dump. But individual actions often need guidance from food distributors and from government. That includes steps such as slightly changing expiration labels, imposing fines for food disposal, or developing connections so that restaurants and grocery stores can divert uneaten food in common-sense ways.

Food waste takes many forms, so there is no one-size-fits-all solution. Instead, we need actions ranging from what individuals and families do right up to local and national government policy changes. The challenge of food waste needs to be approached on a life-course basis: children, including grandchildren, will see what their elders do by avoiding food waste and will develop habits that are more climate-conscious. Small steps yield big gains for climate change.

It's essential to recognize that food waste doesn't just happen in the kitchen. It happens across the whole system of food production. When we say "food waste" we think about food that spoils in the refrigerator or food we don't eat that gets thrown out. But food is also lost during the whole production, storage, and transport from farm to table. Inevitably, estimates differ about just how much food is wasted every year. But the Carbon Almanac Network estimates the amount at over 2 billion tons of food. What does that mean for climate change? Food waste or loss would come to around 8% of greenhouse gas emissions. "If food loss were a country, it would be the third largest greenhouse gas emitter, behind China and the USA."[88] We're damaging our health and our future because of food that is never consumed at all. There's a place where individuals can make a difference.

The Natural Resources Defense Council has estimated that up to 40% of all U.S. food is thrown away uneaten and 87% of that food waste comes from fresh vegetables, fruit, and bread. The Environmental Protection Agency has said that food waste is the biggest component in landfills and incineration. These comprise 24% of such solid waste.

Some changes in this area are surprisingly inexpensive. For example, date labels now are not regulated by federal law and date labels on food are not always reliable. One result of believing date labels is that people throw out perfectly good food. Some leading food producers are now working to promote federal legislation on standardized date labeling.

Changing diet to become a climate-conscious consumer is important, but it's not an all-or-nothing process. Organic or free-range products may well be more expensive than others, and not everyone can afford this step. Reducing meat and dairy consumption is good, too. Becoming a full vegetarian or a vegan is a bigger step and, again, it's not for everyone.

Wendell Berry's work has influenced those concerned with becoming a climate-conscious consumer. Food writer Mark Bittman called Berry the "soul" of the movement, and Michael Pollen designated him its "spiritual father." Eating meat, especially beef, demands that older people, people of all ages, think more deeply about the sources of their food:

> People who eat have a moral responsibility to the sources of their food. People from the city should do an honest, full accounting of the food that they eat. The first thing they'll discover is that they can't do it. They don't know the ecological cost or the cost to the people who did the work of production, what it costs the rural communities.[89]

We know already that big business is connected with environmental destruction. What we may not realize is how much damage is done both by big drug manufacturers and by big food companies. When the incentive for profits is elevated above public health, we will need to look critically at big names like Wal-Mar, GlaxoSmithKline, Tyson Foods, and Monsanto. Whether from fertilizers and agriculture or chemotherapy in medicine, the lines between relates to Big Pharma and Big Food are increasingly blurred.[90]

Regenerative agriculture is much more than organic farming. Regenerative agriculture involves ecologically based practices. These practices might involve no-till farming, rotational grazing, mixed crop rotation, use of cover crops, compost, and manure. Regenerative agriculture means recognizing differences in soil types, water availability, climate, natural surroundings, and biodiversity. Regenerative growers can tailor to site-specific blends of regenerative practices to improve soil health and resilience to climate change.[91] These practices help reintroduce carbon back into the soil and minimize soil disturbance. Regenerative farming is related to the broader project of regeneration as the most effective response to climate change, a project developed by Paul Hawken.[92]

A big problem for regenerative agriculture is that consumers right now don't know how to find food that's reliably produced by regenerative methods. The lack of clear standards leaves the entire domain open to new forms of greenwashing or deception of the public. For example, the U.S. Department of Agriculture (USDA) is now funding what is called "climate-smart" agriculture. The laudable goal is to reduce greenhouse emissions from agriculture, which constitute 10% of total emissions. Tyson Foods is receiving a $61 million grant from USDA to produce something called "climate-smart" beef. Tyson is launching Brazen Beef, the first USDA-approved "climate-friendly" beef. But can we actually define or certify what "climate-smart" beef consists of?[93]

Here is a basic problem with regenerative agriculture. Like organic foods, the goal is compelling. But without clear standards and certification, consumers won't know what they're buying or what they're eating. Hence the necessity for widely understood standards for regenerative agriculture. For example, A Greener World defines regenerative agriculture as "a set of planned agricultural practices that ensure the holding is not depleted by agriculture practices, and over time the soil, water, air and biodiversity are improved or maintained to the greatest extent possible."[94] Other organizations have also come forward with ideas for certification of regenerative foods. But the ambiguous experience of "organic" foods here is instructive. It has taken decades for consumers to have a reasonable expectation of finding and buying organic foods.

We may eventually have success with regenerative agriculture, but some have doubted whether it will be really the key element in climate change: "Despite regenerative agriculture's popularity and its focus on sustainable food production, it fails to tackle systemic social and political issues. As a result, the movement may perpetuate business-as-usual in the food system, rather than transform it."[95] The climate crisis demands much quicker action and may require broader transformation of our entire food system.

8.1   Frances Moore Lappé: Eco-Elder

Figure 8.6 Frances Moore Lappé. Public domain.

Born in 1944, Frances Moore Lappé is renowned for her books, including her 1971 book *Diet for a Small Planet*. The Smithsonian National Museum of American History called that book "one of the most influential political tracts of the times." She has been a proponent of what she terms "Living Democracy," which stands as an essential component of work around climate change. Historian Howard Zinn said about Lappé: "A small number of people in every generation are forerunners, in thought, action, spirit, who swerve past the barriers of greed and power to hold a torch high for the rest of us. Lappé is one of those."

She has also embodied intergenerational climate work in her own family. Lappé and her daughter Anna created the Small Planet Institute to promote education and stronger democracy on a global scale. A climate-conscious diet is one of the major contributions from Frances Moore Lappé, an Eco-Elder for our time.

LOCAL TRANSPORTATION

Transportation is a big part of becoming a climate-conscious consumer. In the USA transportation overall now accounts for the biggest share of emissions: nearly 30% of all greenhouse gas emissions,[96] making it the largest contributor of emissions. It's been getting worse: in three decades before 2020, greenhouse gas emissions in transportation increased more in absolute terms than any other sector.[97] Local transportation is important: 45% of transportation emissions comes from cars, motorcycles, buses, and taxis.[98] Major decarbonizing the economy is possible with more electrification, including EVs.

Electric Vehicles

EVs are battery-operated vehicles. There are now more than 12 million passenger EVs and a million commercial EVs. Since 2015, on a global basis, the proportion of new passenger EVs has been increasing at 50% per year. This trend gives great hope in the future for reducing greenhouse gas emissions from transportation.

The cost of operating an EV will soon be less than a gasoline vehicle, even if the initial purchase price is higher. Apart from climate change, EVs are a superior and less expensive technology.[99] The car industry is quickly electrifying in far-reaching ways.[100]

When it comes to pushing back against climate change, EVs look like a slam-dunk: no more debate about it. True, only 1% of Americans drive EVs, but, in 2022, sales rose by 60%, and California, by 2035, will prohibit buying gas cars at all. However, there has been no change in the proportion of consumers who want to buy an EV just because it's an EV.[101]

The future looks brighter. Bloomberg has estimated that by 2025 EVs will account for nearly a quarter of sales. At least 17 states are thinking about going down the same road. Climate scientists tell us the transport sector is a big producer of greenhouse gas emissions. By aggressively pushing EVs, we could hope to drive down carbon emissions by 80% by mid-century. The 2022 IRA includes lots of incentives to move along the transition from gas to electric cars. What's not to like about EVs?[102]

Alas, there are some shadows along with this bright picture. First of all, buying an EV is not always easy and it's not cheap: in 2023 the average price for an EV had risen to nearly $59,000, about the same as the annual income of a median worker in the USA.[103]

From an environmental standpoint, there are other issues. We need to reflect on what goes into making EVs, where it comes from, and at what human cost:

> While electric vehicles are essential to reducing carbon emissions, their production can exact a significant human and environmental cost. To run, EVs require six times the mineral input, by weight, of conventional vehicles . . . These minerals, including cobalt, nickel, lithium and manganese, are finite resources. And mining and processing them can be harmful for workers, their communities and the local environment.[104]

We need to pay attention to the serious adverse impacts of EVs, including:[105]

- The impact of mining and extraction of lithium, cobalt, nickel, and other metals on people and ecosystems.
- The emissions generated from mining, transporting, assembling, and disposing of EV materials.
- The environmental hazards of processing, recovering, or disposing of all this stuff.

The International Energy Agency tells us that, to move to these EV goals, we'll need more minerals to be mined: a lot more minerals. Some of those minerals come from China. Wherever they come from, there are environmental downsides to mineral mining. Cobalt is one of those key minerals, and taking it out of the ground comes with deforestation, fragmenting habitats, and child labor. Some of these downsides come in central Africa but others in Alaska, where copper and cobalt are available. Similar problems arise for nickel and sodium. Access to minerals is essential for EV batteries. But minerals don't come cost-free.[106]

When we look at all the costs, including hidden costs, for EVs, the picture is more complex:

> Electric cars are drastically cleaner than conventional gasoline vehicles, particularly as renewable power starts to comprise more of the grid.

However, there's a growing awareness that none of this 5,000-pound hardware comes without a carbon cost; there's no such thing as a "zero-emission vehicle," as they have been branded by policymakers and early champions. Steel body panels don't grow on trees. Lithium doesn't just flow out of the ground and into battery plants. And electricity—even the stuff coming from solar panels—isn't captured without a great deal of capital and carbon expense.[107]

Another factor to consider is that the diminished cost for EVs is partly the result of government rebates and tax credits, and these can change:

Fewer new electric vehicles will qualify for a full $7,500 federal tax credit later this year [2023], and many will get only half that, under rules proposed Friday by the U.S. Treasury Department. The rules, required under last year's IRA, are likely to slow consumer acceptance of electric vehicles and could delay President Joe Biden's ambitious goal that half of new passenger vehicles sold in the USA run on electricity by 2030.[108]

In April 2023, EPA put forward two regulatory plans to limit emissions from cars and trucks, but the standards cover only new car sales and the standards don't go into effect until the period 2027 through 2032. The new rules substantially limit the number of EVs that qualify for the $7,500 tax credit. Half a dozen models will now only qualify for $3,750, or half of the original $7,500, and some will get no credit at all.

Should Older People Buy EVs?

One answer, my own personal answer, is simple: no, I don't want a new car and I don't need a new car, so I'm not a customer for an EV or any kind of car at all. Even if cars mostly go electric, it doesn't follow that I should scrap my perfectly functional car and spend money to buy a new car, an EV. The average cost of any new EV is significantly more than a gas-powered car. An EV may be cheaper over a lifespan than a gas-powered car. But for someone 70 or 80 years old, does "cheaper over a lifespan" make the same sense as it does for someone 20 or 30 years old?

In the coming decades (through 2030 and beyond) EVs are clearly becoming more and more popular. We can even predict that (EVs) are the future of personal transportation. At least we can hope so, and, with the IRA and its subsidies for EVs, we can expect that it will happen. Electrification of transportation is essential to avoid a climate change disaster. We can expect that in years to come there will be more charging stations, technology will improve, and competition from big automakers will bring prices down, as the supply chain improves along with battery technology: "Under a new deal, some of Toyota's future hybrid and electric cars will likely be powered by

U.S.-built batteries made from recycled minerals stripped from old Priuses. This move signals the beginning of a potential circular battery supply chain in the USA."[109]

Along with EVs, the number of older Americans is growing, too. The U.S. Census estimates that the proportion of the population over age 65 will grow from 17% in 2020 to 20% by 2030. More older people will mean more people with the ability to buy a new car. So why not buy an EV?

Start with the number of miles people drive each year. According to the Federal High Highway Administration, people over 65 drive an average of 7,600 miles a year, while people aged 20 to 65 average around 15,000. In other words, younger adults drive twice as much as older adults. That's a big difference.

There are also reasons that people of any age have for not buying an EV, including these, each listed here with question marks and the reply from those promoting EVs:

- **Range anxiety?** Will the battery last long enough when I drive the car a longer distance? Actually, most current EV owners driving cars with a range of 250 miles find that to be completely adequate.
- **Few charging stations around?** Most EV owners charge at home.
- **Limited selection and utility?** Since 2015 the EV global marketplace has been growing at a rate of roughly 50% per year. The World Resources Institute estimates that all new cars will be electric by 2040, and California has mandated that for five years before that date. Limited selection soon will not be a problem.
- **Too expensive?** Prices are dropping, and Consumer Reports documents just how much EVs are cheaper than gas-powered cars over their lifetime. Savings over time are bigger than ever and EVs will soon be no more expensive than gas cars[110] (Ewing, 2023). Although they cost more to start with, EVs will soon be a better deal than gas cars. And let's not forget the $7,500 tax credit.[111]
- **Lack of Information?** Lots of people are unfamiliar with EVs, and older people, inevitably, are less familiar than younger people. But older people can learn.

The Real Reason NOT to Buy an EV

Over the course of my own lifetime (I'm 79) the vast majority of greenhouse emissions in the USA have come from cars. Should we simply say "case closed" and expect more older people to buy an EV? Let's look at what some might call a one-cell sample: namely, the author of this book, an elder with sufficient financial assets to buy an EV, as family members have urged me to do. So why don't I do it?

First, a small reason—the simple fact that in a condo, such as the one where I live, it's not physically possible to charge an EV at home. I'm accustomed to visiting gas stations, and an EV charging station isn't far away. So why don't I buy an EV?

Biggest reason: because I don't need one. I don't drive much, I'm retired and I don't commute, and when I do drive, it's not very far away, so range anxiety isn't the problem at all. The problem is that I just don't need a car much and I have perfectly good alternatives. If my car is in a shop for repairs, I can easily find alternatives: borrowing a car from a family member, using a ridesharing service like Uber or Lyft, using public transportation, and so on.

If you ask an amputee why he doesn't buy a bicycle, there's an obvious answer. And I have an answer for not buying an EV, too: I don't need a new car and I'm not buying one any time soon. That answer is not limited to me. There are growing numbers of Americans in the category of the "old-old" (75 to 85) or the "oldest-old" (85-plus). As a generalization, they're not people who are interested in driving a lot, at least not as much as they used to do. For the most part, they're not commuting and today they have more and more options for home delivery of goods. Like me, these older people have limited life expectancy so the argument for "cheaper-over-the-lifespan" for them doesn't make a lot of sense.

There are other issues with EVs, and one of them is that battery technology is changing rapidly. Some auto companies are now claiming that sodium-based batteries will soon be a better choice. Does that mean waiting to buy an EV until battery technology improves? It may be that you can wait forever as technology always improves. If you're driving an older EV or hybrid with lesser tech, but adequate range, it may be that you should just keep doing what you're now doing

One could argue that, as a climate advocate myself, I should buy and drive an EV anyway, because it sets a good example for others. That argument sounds plausible, but it overlooks the opportunity cost factor. The average price of an EV in 2023 was around $65,000, and prices are not going down, making it harder to buy an EV.[112] $65,000 could go a long way in contributions to nonprofit climate advocacy or for many other good causes. If I don't drive much at all, and if, like others over 65, my miles driven are lower than younger people, then it's hard to see why it's setting a good example for an EV vehicle spending most of its time in a garage.

The other EV problem is that most policy proposals focus on new cars rather than older ones. If I'm not planning to buy a new car (I'm not), then I just keep on polluting, as with my still-functioning methane gas heater. Policies are proposed to speed up adoption of EVs[113] and that's clearly a good thing.[114]

If we were in any doubt about being a reluctant EV car buyer, the best reply comes from another one-call study: Nobel-prize-winning economist Paul Krugman[115] who said this:

> When I won the Nobel Prize in 2008, Princeton quickly set up a special event on campus and reserved a parking space for me in front of

Robertson Hall. But when I drove up in my 2004 Jetta, the security people frantically tried to wave me away. They clearly didn't find it plausible that a laureate would be driving such a modest car.

Paul Krugman wasn't about to buy a new car just because he suddenly came into a lot of money. Whether you're a world-famous economist (like Krugman) or a semi-retired gerontologist (like me), there's a place for people who don't want to be big-time consumers of cars. Sometimes being a climate-conscious consumer means not being a consumer at all.

Public Transportation

What if the debate about buying EVs is simply the wrong debate? Should your next car be an EV, or maybe no car at all?[116] Debates about EV and climate mitigation are often misconceived. Even worse, there are times when EVs get dragged into culture wars, and consumer choice about EVs gets politicized. There are voices claiming that the government will take away your gas stove and stop you from eating hamburgers, and, yes, voices that claim environmentalists want to take your car away.[117] Others are concerned about the energy grid and where all the electricity comes from. Even under the best of circumstances, we have to be cautious about believing that EVs will save us from climate change. As one climate advocate put it to me: "I don't know why people keep talking about how electric cars are going to save us. Makes me crazy. It especially makes me crazy when one considers we haven't figured out how to make electricity without burning fossil fuels!"

The problem is a false dichotomy: EITHER we continue to drive gasoline-powered cars, mostly the current option, OR we go to EVs and try to deal with issues of implementation: purchase price, access to charging stations, and so on. But what if this is the wrong set of alternatives? We might say, no, forget individual automobiles and go for public transportation. But public transit, whether rail or bus, poses other problems: for example, the cost of installation and locating routes in rural or low population density areas. Still another alternative would be shared individual vehicles, such as Uber provides. Beyond the private shared vehicle approach, there are low-cost or nonprofit approaches to shared individual vehicles, such as what ITN America does in many places around the United States. It is certainly possible to require that shared vehicles be EVs, whether owned by for-profit or not-for-profit organizations.

The point about enhancing public transportation and alternate motor transportation is crucial.[118] But public transit is in trouble. In 2019 only 5% of workers commuted by public transit in 2019 and ridership declined during the coronavirus pandemic. Think about the question: "Should your next car be electric—or no car at all?" If the road to decarbonization means getting the incentives right, then maybe the last thing we need to do is to subsidize more cars, whether they are electric or not. Pushing for EVs also overlooks a big question: what kind of EV are we talking about? More and more are SUVs, which can cancel out gains from shifting out of fossil into electricity.[119] As Elizabeth Kolbert notes, the world is moving toward heavier cars at a time when it should be doing precisely the reverse.

We can look around the world and see that it doesn't have to be this way.[120] For example, when it comes to public transportation Europe is very different from the USA. European countries generally have much more extensive train network than America. They offer easy intercity travel without relying on carbon-consuming air transport or cars. Public transportation systems in Europe are not limited to trains but include buses, trams, and bike-sharing programs. In short, you're far more likely to find eco-friendly travel in Europe than in the USA.

The point is to see that it is a false dichotomy to frame the challenge as either current gas-powered cars or EVs. That way of framing the challenge describes it entirely in terms of physical technology in motor vehicles. When Lyft and Uber emerged, that constituted a new kind of social technology: well, not entirely "new" because taxi cabs have been around for a long time. We need to acknowledge that the advent of EVs, even if adopted on a massive scale, may not be the solution to climate change.[121]

This discussion has particular importance for climate change in an aging society because as people get older, they face specific age-related challenges. For example, driving itself may be more difficult because of sense impairments (vision and hearing). At the same time, mobility problems may make it more difficult to walk to a bus station or a subway station for public transit. For medical or other professional appointments—more common in old age—flexibility of individual car transportation turns out to be important. All these considerations make it important to think about how to provide local transportation to older people and also reduce greenhouse emissions. Local transportation for climate-conscious consumers is not a challenge they can solve on their own. It is unrealistic and unreasonable to expect that simply trying to sell more EVs to elders will solve the problem.

The boom in EVs could give a reason for optimism: for example, in Norway now EVs constitute four out of every five new vehicles sold. In 2016, that figure was only one out of five. But David Wallace-Wells identifies the fly in the ointment here: EVs now on the road are perhaps 2% of the global fleet, which is very far from 100%, where it could make a difference in global warming. In the USA, EVs are only 1% of the 279 million vehicles on the road. There are transit alternatives, too. In 2020 Americans bought twice as many e-bikes as EVs.[122]

Enthusiasts for EVs need to keep this perspective in mind. The cost of EVs is just too high for many people and is likely to remain so.[123] For the time being, I'll just keep my car (mostly in the garage) and feel guilty whenever I drive it. What's the old saying? "Guilt is the gift that keeps on giving."

8.2 Lester Brown: Eco-Elder

Figure 8.7 Lester R. Brown. Public domain.

Lester R. Brown could be described as the great hero of contemporary environmental advocacy. He has been the founder of the Worldwatch Institute and the Earth Policy Institute and is the author or co-author of over 50 books on global environmental issues, including *World on the Edge* and *Full Planet, Empty Plates*.

In recent years his attention has been turned to climate change and how we can respond to it. In 1978, in *The Twenty-Ninth Day*, Brown was already warning of "the various dangers arising out of our man-handling of nature . . . by overfishing the oceans, stripping the forests, turning land into desert."

At age 79 he said: "I'm often asked, am I an optimist or a pessimist? I like to think I'm a realist. We're going to have to move very fast if we want to prevent climate change from spiraling out of control."[124] Two years after that statement he published *The Great Transition: Shifting from Fossil Fuels to Solar and Wind Energy*.

Brown has been described as a Jeremiad prophet of doom, yet in old age he began speaking in optimistic terms. *Scientific American* published a biography of him under the title of "Persistent Prophet."

Lester Brown is the premier model of an Eco-Elder, an example for us all.

Notes

1 C40 Cities, "New Research Shows How Urban Consumption Drives Global Emissions" (June 12, 2019) at: www.c40.org/news/new-research-shows-how-urban-consumption-drives-global-emissions/

2 For an in-depth account of the growth of consumer culture in the 20th century, see Kerryn Higgs' book *Collision Course: Endless Growth on a Finite Planet* (MIT Press, 2014). See also Sandra Goldmark, *Fixation: How to Have Stuff without Breaking the Planet* (Island Press, 2020).

3 From Union of Concerned Scientists, "Cooler Smarter: Geek Out on the Data!" at: www.ucsusa.org/resources/cooler-smarter-geek-out-data What we don't consider in this chapter is air travel. Intercontinental air travel is among the most carbon-intensive activities that exist, and there's very little technological substitute for it. Electric air travel is possible but, for most of us, over the horizon. Domestic air travel is important, of course. With the advent of Zoom and other means of remote work or conferencing, many varieties of travel will be reconsidered with climate considerations a factor. But, for reasons of space, in this chapter it is not considered, except to recognize the importance of air travel and inequality, discussed in Chapter 6, Who Is to Blame for Global Warming? The top 1% group flying in private jet planes generates 14 times more greenhouse gas emissions per passenger mile than people flying on commercial flights. Apart from air travel, the 1% flying in private jets generate 50 times more than those in trains, as much as millions of private cars. Are the top 1% in private jets paying their way? No, they're not. Private jets make up 16% of the flights handled by the air traffic control system, but they contribute only 2% of funding for that system. That's why legislators have introduced what they're calling the FATCAT Act, which stands for Fueling Alternative Transportation with a Carbon Aviation Tax. Appropriately named. The money from that new technology could go toward air quality monitoring and more investment in public transit, which generates far less greenhouse gas emissions. See L. Robert Reich, "Inequality Media" (August 6, 2023).

4 See "Why Do We Keep Buying New Stuff?" at: https://eu.patagonia.com/nl/en/stories/why-do-we-keep-buying-new-stuff/story-144207.html

5 See "Can Reducing Consumerism Help Us Save the Planet and Find Happiness?" A Conversation with J.B. MacKinnon, from *The Way Forward Regenerative Conversations* at: https://regenerationthewayforward.podbean.com/ See J.B. MacKinnoon, *The Day the World Stops Shopping: How Ending Consumerism Saves the Environment and Ourselves* (Ecco, 2021).

6 Tatianna Schlossberg, *Inconspicuous Consumption: The Environmental Impact You Don't Know You Have* (Grand Central Publishing, 2019), p. 3. See: "7 Sustainable Living Tips That Prove Small Changes Can Make a Big Difference" *Brightly* (May 8, 2023) at: https://brightly.eco/blog/sustainable-living-tips See also: https://brightly.eco/eco-advice

7 H.R. Moody, "From Successful Aging to Conscious Aging" in *Successful Aging through the Lifespan*, edited by May Wykle, Peter Whitehouse and Diana Morris (Springer, 2004). See also H.R. Moody, "Critical Theory and Critical Gerontology" in *Encyclopedia of Aging* (Springer, 1995).

8 Juliet Schor, *Plenitude: The New Economics of Truth Wealth* (Penguin, 2010).

9 See Aimee Drolet, Norbert Schwarz and Carolyn Yoon (eds.), *The Aging Consumer: Perspectives from Psychology and Economics* (Routledge, 2009).

10 Simon Mair, "Why Strike Action Is Climate Action" *The Conversation* (Apr. 25, 2023) at: https://theconversation.com/why-strike-action-is-climate-action-203815

11 Daniela Blei, "Politics, Values, and the Green Transition" *Stanford Social Innovation Review* (Summer 2023) at: https://ssir.org/articles/entry/politics_values_and_the_green_transition

12 Michael Coren, "Climate Coach" at *The Washington Post* gives helpful guidance on becoming a climate-conscious consumer: "Why You Should Buy Everything Used" *Washington Post* (May 23, 2023) at: www.washingtonpost.com/climate-environment/2023/05/23/buy-resale-store-second-hand-clothes-furniture/

13 Bill McKibben, "From Climate Exhortation to Climate Execution" *The New Yorker* (Dec. 27, 2022) at: www.newyorker.com/news/daily-comment/from-climate-exhortation-to-climate-execution

14 ReFed, "Roadmap to 2030: Reducing U.S. Food Waste by 50%" (2022) at: https://refed.org/food-waste/the-problem/ To put it in more tangible terms: An area larger than Canada is needed to grow food that's never eaten.

15 "Food Pyramid, Compass, or Plate? Food Guides Can Help Improve Health and Address Climate Change" at: https://sustainableamerica.org/blog/food-pyramid-compass-or-plate-food-guides-can-help-improve-health-and-address-climate-change/ For action steps, see: "How Our Food System Affects Climate Change" at: https://foodprint.org/issues/how-our-food-system-affects-climate-change/ For more action steps, see "Will Climate Cookbooks Change How We Eat?" *Grist* (Nov. 20, 2023) at: https://grist.org/culture/climate-cookbooks-sustainable-eating-low-waste-kitchen/ See also Michael Hoffman, *Our Changing Menu: Climate Change and the Foods We Love and Need* (Cornell University Press, 2021). For more on food and climate, see: https://climateaging.bctr.cornell.edu/post/how-the-intersection-of-climate-change-and-food-can-inspire-change and www.ourchangingmenu.com/

16 "Global Greenhouse Gas Emissions by Sector" *The Carbon Almanac* at: https://thecarbonalmanac.org/013/

17 Even if recycling seems simple, it can be remarkable cost-effective: "Curbside Recycling Turns Out to Be a Surprisingly Good Climate Investment" *Anthropocene* (May 30, 2023) at: www.anthropocenemagazine.org/2023/05/curbside-recycling-turns-out-to-be-a-surprisingly-good-climate-investment/

18 Mary Meade, "10 Ways You Can Fight Climate Change" from "Your Green Life" *Green America* (Fall 2019) at: www.greenamerica.org/magazine/your-green-life-2019

19 Green America has created its National Green Pages to identify find the climate-conscious products and services from socially responsible businesses. They also offer Green Business Certification, so consumers can be assured they are becoming climate-conscious consumers. For more on this resource, see: www.greenamerica.org/green-businesses-products-services

20 Grist, "5 Ways Climate Change Made Life More Expensive in 2022" *Grist* (Dec. 21, 2022) at: https://grist.org/economics/5-ways-climate-change-made-life-more-expensive-in-2022/?utm_source=newsletter&utm_medium=email&utm_campaign=beacon

21 On "nudges" and incremental choice, see Richard Thaler and Cass Sunstein, *Nudge: Improving Decisions about Health, Wealth, and Happiness* (Penguin, 2009). See: "How to Nudge Americans to Reduce Their Housing Exposure to Climate Risks" at: www.brookings.edu/articles/how-to-nudge-americans-to-reduce-their-housing-exposure-to-climate-risks/ We've already looked in an earlier chapter on the impact of insurance and choices. See also: "How Climate Change Could Cause a Home Insurance Meltdown" at: www.npr.org/2023/07/22/1186540332/how-climate-change-could-cause-a-home-insurance-meltdown

22 Grantham Institute, "9 Things You Can Do about Climate Change" Imperial College, London (2023) at: www.imperial.ac.uk/stories/climate-action

23 Umair Irfan, "Consumers, Not Corporations, Saved the Power Grid. What Else Can We Do?" *Vox* (Oct. 31, 2022) at: www.vox.com/science-and-health/23340991/power-demand-response-blackout-consumer-climate-change-california-texas-cop27

24 Practical solutions are available: "ENERGY STAR Home Upgrade: Six Energy Efficiency Improvements" at: www.energystar.gov/products/energy_star_home_upgrade

25 https://climatescience.org/advanced-personal-action
26 Rocky Mountain Institute, "All-Electric Homes: A Health Professional's Guide" (2023) at: https://rmi.org/all-electric-homes-a-health-professionals-guide/
27 *Carbon Almanac Network*, Aug. 24, 2023.
28 Alison F. Takemura, "Electrifying Your Home Can Be a Huge Hassle" *Canary Media* (Mar. 8, 2023) at: www.canarymedia.com/articles/electrification/electrifying-your-home-can-be-a-huge-hassle-helio-home-wants-to-help
29 Ula Chrobak, "How Heat Pumps Can Help Fight Global Warming" *Popular Science* (Mar. 3, 2020) at: www.popsci.com/story/environment/heat-pumps-emissions-climate-change/
30 See Rewiring America at: www.rewiringamerica.org/app/ira-calculator
31 Ajit Niranjan, "'You Can Walk Around in a T-Shirt': How Norway Brought Heat Pumps in from the Cold" at: www.theguardian.com/environment/2023/nov/23/norway-heat-pumps-cold-heating
32 For guidance on heat pumps, see www.energy.gov/energysaver/programmable-thermostats This is only one example, and there are many more. Earth Hero is a volunteer-driven nonprofit phone app that can offer concrete, practical ways reduce our impact on the planet: https://apps.apple.com/ca/story/id1613364866
33 Bronwyn Adcock, "Electric Monaros and Hotted-Up Skateboards: The 'Genius' Who Wants to Electrify Our World" *The Guardian* (Feb. 4, 2022) at: www.theguardian.com/australia-news/2022/feb/05/electric-monaros-and-hotted-up-skateboards-the-genius-who-wants-to-electrify-our-world
34 Dori Newman, "How Does the Inflation Reduction Act Affect Households?" *Clear Energy Finance Forum* (Apr. 17, 2023), Yale Center for Business and the Environment
35 David Roberts, "Building a Movement That Can Take Full Advantage of the IRA" *Volts* (Apr. 26, 2023) at: www.volts.wtf/p/building-a-movement-that-can-take
36 These figures are from an actual example by Robert Whitehair, "The Benefits and Ease of Electrification" *The Daily Journal* (Apr. 4, 2023) at: www.smdailyjournal.com/opinion/guest_perspectives/the-benefits-and-ease-of-electrification/article_ac910fc6-d293-11ed-9604-6b2f2829135c.html For more on Whitehair's story, see: www.youtube.com/watch?v=GRu_FwmidlU&t=10s
37 Union of Concerned Scientists, "Our Overdependence on Methane Gas Is Costly: We Need Policymakers to Pass Clean Energy Legislation Now" (May 31, 2022) at: https://blog.ucsusa.org/ashtin-massie/our-overdependence-on-methane-gas-is-costly-we-need-policymakers-to-pass-clean-energy-legislation-now/
38 Paul Hope, "Pros and Cons of Induction Cooktops and Ranges" *Consumer Reports* (Jan. 13, 2023) at: www.consumerreports.org/electric-induction-ranges/pros-and-cons-of-induction-cooktops-and-ranges-a5854942923/
39 Mary Farrell and Paul Hope, "CR's Complete Guide to Induction Cooking" *Consumer Reports* (Oct. 11, 2022) at: www.consumerreports.org/appliances/ranges/guide-to-induction-cooking-a2539860135/#:~:text=According%20to%20CR's%20June%202022,induction%20appliance%20in%20their%20kitchen.
40 Yale Climate Connections, "Can Induction Stoves Convince Home Cooks to Give Up Gas?" at: https://yaleclimateconnections.org/2023/01/can-induction-stoves-convince-home-cooks-to-give-up-gas/?utm_source=Weekly+News+from+Yale+Climate+Connections&utm_campaign=0160cd15f7-EMAIL_CAMPAIGN_2023_01_12_09_15&utm_medium=email&utm_term=0_-0160cd15f7-%5BLIST_EMAIL_ID%5D
41 Paul Hope, "Induction Cooktops and Ranges Are So Good You May Not Miss Your Gas Appliance" *Consumer Reports* (Jan. 12, 2023) at: www.consumerreports.org/appliances/ranges/induction-cooktops-ranges-are-so-good-you-may-not-miss-gas-a8912134554/?utm_source=substack&utm_medium=email   See also: Ginia Bellafante, "No One Is Coming for Your Gas Stove Anytime Soon" *NYTimes* (Jan. 21, 2023) at: www.nytimes.com/2023/01/21/nyregion/gas-stoves-nyc.html?smid=nytcore-ios-share&referringSource=articleShare

42 Jeff Brady, "Gas Stoves Leak Climate-Warming Methane Even When They're Off" *National Public Radio* (Jan. 27, 2022) at: www.npr.org/2022/01/27/1075874473/gas-stoves-climate-change-leak-methane

43 International Energy Agency, Global Methane Tracker 2022, IEA, Paris at: www.iea.org/reports/global-methane-tracker-2022

44 Spencer Bokat-Lindell, "Is the Era of Gas Stoves Burning Out?" *NY Times* (Jan. 25, 2023) at: www.nytimes.com/2023/01/25/opinion/gas-stove-ban-induction.html?

45 Victor St. Martin, "How Gas Stoves Became Part of America's Raging Culture Wars" *Inside Climate News* (Jan. 22, 2023) at: https://insideclimatenews.org/news/22012023/gas-stoves-culture-wars-health-concerns/ See also: Greg Dalton, "Amid a Culture War Over Gas Stoves, Research on Health Impacts Speaks for Itself" *Climate One* (Jan. 20, 2023).

46 Tyler Wells Lynch, "The Best Slide-In Electric Ranges" *NY Times Wirecutter* (Jan. 13, 2023) at: www.nytimes.com/wirecutter/reviews/best-slide-in-electric-ranges/

47 www.energy.gov/energysaver/programmable-thermostats See also: "Be an Earth Hero" at: https://apps.apple.com/ca/story/id1613364866

48 Pema Chodron, *Start Where You Are: A Guide to Compassionate Living* (Shambhala, 2018).

49 The title of this section of the chapter of course is the title of the ground-breaking 1971 book by Frances Moore Lappé. Hers was the first widely sold book pointing to the role of plant protein instead of meat production. I was myself 26 years old at the time. If those, like me, who are elders today had gotten the message earlier in life, who can say how much health damage would have been prevented, not to mention damage to the environment? But, as the message on this page says: you are not too old and it is not too late: the indispensable point for climate change in an aging society.

50 George Monbiot, "How What We Eat and How We Eat It Helps Contribute to Climate Change" *Literary Hub* (Aug. 8, 2022) at: https://lithub.com/how-what-we-eat-and-how-we-eat-it-helps-contribute-to-climate-change/

51 Climate and Environment, United Nations (Mar. 9, 2021) at: https://news.un.org/en/story/2021/03/1086822

52 *The Conversation*, "Scrapping Use-By Dates Could Prevent Huge Amounts of Food Waste—Here's What Else Could Help" https://theconversation.com/scrapping-use-by-dates-could-prevent-huge-amounts-of-food-waste-heres-what-else-could-help-188085 See also Linnea Harris, "Is It Really Expired? The Truth About Food 'Expiration' Dates" *EcoWatch* (Aug. 11, 2023) at: www.ecowatch.com/food-expiration-labels-dates-safety-ecowatch.html

53 *Foodprint*, "A,B,C's of Food Waste" (2023) at: https://foodprint.org/wp-content/uploads/2021/05/2021_05_03_FP_ABCsOfReducingFoodWaste_FINAL.pdf For simple steps to stop food waste and fight climate change see also: "Fight climate change by preventing food waste" at: www.worldwildlife.org/stories/fight-climate-change-by-preventing-food-waste See also Kathleen Ronayne, "Food Waste Becomes California's Newest Climate Change Target" *Associated Press* (Dec. 8, 2021) at: www.yahoo.com/news/food-waste-becomes-californias-newest-050449405.html See also Martin Igini, "10 Food Waste Statistics in America" *Earth.Org* (Nov. 23, 2022) at: https://earth.org/food-waste-in-america/

54 For more on the Sustainable Food Trust see: https://sustainablefoodtrust.org/

55 I note the contribution of Prof. Jennifer Sasser in making these points about the limits on individual consumer choice.

56 Alexandra Applegate, "How to Choose Climate-Friendly Foods If You Care About the Environment" *BuzzFeed News* (Sept. 4, 2022).

57 For more on intersectionality and vulnerable groups, see Leah Thomas, also known as Green Girl Leah, *The Intersectional Environmentalist: How to Dismantle Systems of Oppression to Protect People + Planet* (Voracious, 2022).

58 Abrahm Lustgarten et al., "What You Need to Know About the Water Crisis in the West" *ProPublica* (May 27, 2015) at: https://projects.propublica.org/killing-the-colorado/story/what-you-need-to-know/

59 Kenny Torella, "The Colorado River Is Going Dry . . . to Feed Cows" *Vox* (Apr. 19, 2023) at: www.vox.com/the-highlight/23655640/colorado-river-water-alfalfa-dairy-beef-meat See also: Brian Richter, et al., "Water Scarcity and Fish Imperilment Driven by Beef Production" *Nature Sustainability* (Apr., 2020) at: www.nature.com/articles/s41893-020-0483-z.epdf See also Brian Kateman, "The Meat and Dairy Industries Peddle Misinformation. It's Time for Plant-Based Food Brands to Fight Back" *Fast Company* (May 12, 2023).

60 Amanda Schupak, "Climate-Friendly Diets Can Make a Huge Difference—Even If You Don't Go All-Out Vegan" *The Guardian* (June 4, 2022) at: www.theguardian.com/environment/2022/jun/04/meat-diets-climate-emissions-plant-based-vegan See also Daniel Oropeza, "These Simple Food Choices Can Drastically Reduce Your Environmental Impact" *LifeHacker* (May 22, 2023) at: https://lifehacker.com/these-simple-food-choices-can-drastically-reduce-your-e-1850456028?utm_source=pocket-newtab

61 Physician Committee, "A Vegan Diet: Eating for the Environment" (2021) at: www.pcrm.org/good-nutrition/vegan-diet-environment

62 Stephanie Studer, "Which Type of Plant-Based Milk Is Best?" *The Economist* (Nov. 5, 2021).

63 See "Eating Our Way to Extinction" narrated by Kate Winslet.

64 Ezra Klein, "The Hidden Costs of Cheap Meat" *NY Times* (Nov. 29, 2022) at: www.nytimes.com/2022/11/29/opinion/ezra-klein-podcast-leah-garces.html

65 Tatianna Schlossberg, *Inconspicuous Consumption: The Environmental Impact You Don't Know You Have* (Grand Central Publishing, 2019), p. 90.

66 Libby Leonard, "Regenerative Grazing 101: Everything You Need to Know" *EcoWatch* (Apr. 7, 2023) at: www.ecowatch.com/regenerative-grazing-facts-ecowatch.html

67 Faith Ann, "Going Vegan Isn't an Option for Everyone" (Jan. 2, 2020) at: https://aninjusticemag.com/going-vegan-isnt-an-option-for-everyone-53ad0069136f

68 Emma Bryce, "Which Diet Is More Climate Friendly: Novel Foods or Mostly Vegan?" *Anthropocene* (Apr. 29, 2022) at: www.anthropocenemagazine.org/2022/04/which-diet-is-more-climate-friendly-novel-foods-or-mostly-vegan/

69 Emma Bryce, "Food Apps Have Untapped Potential to Nudge People Toward Greener Diets" *Anthropocene Magazine* (May 6, 2022) at: www.anthropocenemagazine.org/2022/05/online-food-apps-have-untapped-potential-to-nudge-consumer-diets/ See also: Sander van der Linden, "What's the Best Way to Shrink Your Carbon Footprint?" at: www.nytimes.com/interactive/2022/12/15/opinion/how-reduce-carbon-footprint-climate-change.html?campaign_id=39&emc=edit_ty_20221216&instance_id=80337&nl=opinion-today&regi_id=20387967&segment_id=120013&te=1&user_id=19f5d92b505a29f7757bf318ae75163e

70 Emma Bryce, "A 15-Year Snapshot of US Diets Reveals a Gradual Shift Away from Beef" *Anthropocene* (May 22, 2022) at: www.anthropocenemagazine.org/2022/05/a-15-year-snapshot-of-us-diets-reveals-a-gradual-shift-away-from-beef/

71 Bryce Emma, "Reducing Food Waste Is an Overlooked Solution to Saving Endangered Species" *Anthropocene* (Apr., 2022) at: www.anthropocenemagazine.org/2022/04/reducing-food-waste-is-an-overlooked-solution-to-saving-endangered-species/

72 Emma Bryce, "Not All Plant-Based Diets Are Equal . . . for Health or the Environment" *Anthropocene* (Nov. 18, 2022) at: www.anthropocenemagazine.org/2022/11/not-all-plant-based-diets-are-equal-for-health-or-the-environment/

73 "The 57,000 Most-Commonly Consumed Foods, Ranked by Environmental Impact" at: www.anthropocenemagazine.org/2022/08/the-57000-most-commonly-consumed-products-ranked-by-environmental-impact/

74 https://greenly.earth/en-us/blog/ecology-news/global-food-waste-in-2022

75 "Some Numbers About Food Waste" at: https://ivaluefood.com/ See also: "Tips to Reduce Food Waste" at: www.nytimes.com/2015/03/04/dining/tips-to-reduce-food-waste.html

76 Max Kim, "South Korea Has Almost Zero Food Waste. Here's What the US Can Learn" *The Guardian* (Nov. 20, 2022) at: www.theguardian.com/environment/2022/nov/20/south-korea-zero-food-waste-composting-system?utm_source=pocket-newtab

77 Grundig: Respect Food at: www.respectfood.com/article/11-facts-about-food-wastage/

78 Libby Leonard, "Food Waste 101: The Facts and Solutions" *EcoWatch* (Nov. 22, 2022) www.ecowatch.com/food-waste-guide.html

79 Rachael Ajmera, "10 Realistic Ways to Eat Less Processed Food" *HealthLine* (June 22, 202) at: www.healthline.com/nutrition/how-to-eat-less-processed-food#8.-Try-some-simple-food-swaps

80 Max Graham, "A Simple Way to Prevent Heaps of Methane Pollution: Composting" *Grist* (May 15, 2023) at: https://grist.org/food/food-waste-prevent-methane-pollution-compost/

81 Grundig: Respect Food at: www.respectfood.com/article/what-the-difference-is-between-use-by-sell-by-and-best-before-dates/ See also Yasmin Tayag, "Expiration Dates Are Meaningless" *Atlantic Monthly* (Nov. 30, 2022) at: www.theatlantic.com/health/archive/2022/11/expiration-dates-food-waste-safety/672311/?utm_source=pocket-newtab

82 It comes down to smart shopping and smart living: don't buy more than you need; don't be deceived by expiration dates; and compose when you can. In short, there are many ways to reduce, reuse, and recycle food that otherwise would become waste.

83 See "The Huge Climate Impact of the Middle Classes" at: www.theguardian.com/environment/2023/nov/20/revealed-huge-climate-impact-of-the-middle-classes-carbon-divide

84 "Why It's Okay to Ignore Food Expiration Dates" *Washington Post* (May 18, 2023) at: www.washingtonpost.com/climate-solutions/2023/05/17/food-expiration-dates-best-by/

85 Carbon Almanac Network, "Food Waste Action Week" (Mar. 6, 2023) at: https://thecarbonalmanac.org/food-waste-action-week/ See also: Earth.org at: https://earth.org/what-is-food-waste/

86 Lots of Food Gets Tossed. These Apps Let You Buy It, Cheap. Several companies say they are tackling food waste by connecting people with unsold food from restaurants and grocery stores. www.nytimes.com/2022/09/20/climate/food-waste-app.html

87 Alexandra Canal, "Americans Discard $1,600 in Food Per Year—How One Brand Is Fighting Food Waste" *Yahoo Finance* (May 13, 2022).

88 Earth.org, 2023.

89 Hope Reese, "A Champion of the Unplugged, Earth-Conscious Life, Wendell Berry Is Still Ahead of Us" *Vox* (Oct. 9, 2019) at: www.vox.com/the-highlight/2019/10/2/20862854/wendell-berry-climate-change-port-royal-michael-pollan

90 Stan Cox, *Sick Planet: Corporate Food and Medicine* (Pluto Press, 2008).

91 Natural Resource Defense Council, "Regenerative Agriculture Part 3: The Practices" (Jan. 15 2021) at: www.nrdc.org/bio/arohi-sharma/regenerative-agriculture-part-3-practices See also: Danielle Arigoni, *Climate Resilience for an Aging Nation* (Island Press, 2023).

92   Paul Hawken, *Regeneration: Ending the Climate Crisis in One Generation* (Penguin, 2021). See also: https://regeneration.org/

93   Arielle Samuelson, "The Mystery of Climate-Friendly Beef" *Heated* (May 25, 2023) at: https://heated.world/p/the-mystery-of-climate-friendly-beef

94   A Greener World at: https://agreenerworld.org/certifications/certified-regenerative/

95   Anja Bless, " 'Regenerative Agriculture' Is All the Rage But It's Not Going to Fix Our Food System" *Pelican Web* (June, 2023) at: www.pelicanweb.org/solisust-v19n06page22.html

96   EPA, "Fast Facts on Transportation Greenhouse Gas Emissions" at: www.epa.gov/greenvehicles/fast-facts-transportation-greenhouse-gas-emissions

97   *Ibid:*www.epa.gov/greenvehicles/fast-facts-transportation-greenhouse-gas-emissions

98   https://ourworldindata.org/co2-emissions-from-transport In this discussion attention is limited to local transportation, excluding aviation, a domain that is generally beyond the scope of individual agency. Questions of technology and public policy are important around aviation. For more on aviation and becoming a climate-conscious consumer, see Stefan Gossling and Paul Upham (eds.), *Climate Change and Aviation: Issues, Challenges and Solutions* (Routledge, 2009); and Christopher de Bellaigue, *Flying Green: On the Frontiers of New Aviation* (Columbia Global Reports, 2023).

99   Bridget Morawski, "Electric Vehicles 101: Everything You Need to Know" *EcoWatch* (Dec. 22, 2022) at: www.ecowatch.com/electric-vehicles-guide-2655917104.html

100 "The Future Lies with Electric Vehicles" *The Economist* (Apr. 14, 2023) at: www.economist.com/special-report/2023/04/14/an-electric-shock See also Tom Randall, "Electric Cars Pass a Crucial Tipping Point in 23 Countries Bloomberg" (Aug. 27, 2023). Once 5% of new car sales go fully electric, everything changes—according to a Bloomberg Green analysis of the latest EV adoption curves: www.bloomberg.com/news/articles/2023-08-28/electric-cars-pass-a-crucial-tipping-point-in-23-countries

101 "EV Market Share Is Growing Because the Vehicles Keep Getting Better" at: https://insideclimatenews.org/news/08062023/inside-clean-energy-electric-vehicle-market-features/

102 Plug In America, "Top 10 Reasons I Like Our Electric Car" (Aug. 11, 2016) at: https://pluginamerica.org/story/jim-t-top-10-reasons/

103 Kelley Blue Book at: https://mediaroom.kbb.com/2023-04-11-After-Nearly-Two-Years,-New-Vehicle-Transaction-Prices-Fall-Below-Sticker-Price-in-March-2023,-According-to-New-Data-from-Kelley-Blue-Book See also Michael Coren, "Buy an Electric Vehicle Now or Wait? Here's How to Decide" *Washington Post* (Apr. 11, 2023) at: www.washingtonpost.com/climate-environment/2023/04/11/electric-vehicle-buying-guide

104 Aaron Steckelberg, et al., "The Underbelly of Electric Vehicles" *Washington Post* (Apr. 27, 2023) at: www.washingtonpost.com/world/interactive/2023/electric-car-batteries-geography/

105 Michael Coren, "The Climate Coach" *Washington Post* (May 9, 2023) at: www.washingtonpost.com/climate-environment/2023/05/09/climate-change-solutions-vegan-electric-car-batteries/

106 Tim Lydon, "The EV Revolution Brings Environmental Uncertainty at Every Turn" *The Revelator: EcoWatch* (Jan. 15, 2023) at: www.ecowatch.com/electric-vehicles-environment.html See also: "The U.S. Needs Minerals for Electric Cars. Everyone Else Wants Them Too" *NY Times* (May 21, 2023) at: www.nytimes.com/2023/05/21/business/economy/minerals-electric-cars-batteries.html

107 Bloomberg Green, "Bloomberg Green's Electric Car Ratings" (July 6, 2022) at: www.bloomberg.com/graphics/electric-vehicles/

108 Tom Krisher, Fatima Hussein and Matthew Daly, "Many Electric Vehicles to Lose Big Tax Credit with New Rules" *AP News* (Mar. 31, 2023) at: https://apnews.com/article/electric-vehicles-tax-credit-7500-c562cb2d3509e93dc81d3b7d395725af

109 From *Chop Wood, Carry Water* a newsletter highlighting good news you never heard about: https://chopwoodcarrywaterdailyactions.substack.com/

110 Jack Ewing, "Electric Vehicles Could Match Gasoline Cars on Price This Year" *New York Times* (Feb. 10, 2023) at: www.nytimes.com/2023/02/10/business/electric-vehicles-price-cost.html?smid=nytcore-ios-share&referringSource=articleShare

111 James Pothen, "Inside Clean Energy: Some EVs Now Pay for Themselves in a Year" *Inside Climate News* (July 14, 2022) at: https://insideclimatenews.org/news/14072022/inside-clean-energy-electric-vehicles-gas-prices/

112 "Spending on new cars by the lowest 20 percent of earners dropped to its lowest level in 11 years. Meanwhile, spending on new cars by the top 20 percent reached its highest level on record, going back to 1984, according to the most recent data from the 2021 Consumer Expenditure Survey, not adjusted for inflation . . . In 2017. . . there were 11 (EV) models available on the U.S. market for less than $20,000 . . . By the end of 2022, there were four. Then, by March 2023, only 2." From Rachel Siegel and Jeanne Whalen, "New cars, once part of the American dream, now out of reach for many" Washington Post (May 7, 2023).

113 "E.P.A. Lays Out Rules to Turbocharge Sales of Electric Cars and Trucks" The Biden administration is proposing rules to ensure that two-thirds of new cars and a quarter of new heavy trucks sold in the United States by 2032 are all-electric. *NY Times* (Apr. 12, 2023) at: www.nytimes.com/2023/04/12/climate/biden-electric-cars-epa.html

114 "Editorial: EPA Wants to Speed up EV Switch. Good, the Planet Needs It" *Los Angeles Times* (Apr. 12, 2023) at: www.latimes.com/opinion/story/2023-04-12/editorial-the-epa-proposes-a-speedy-switch-to-electric-vehicles-the-planet-needs-it

115 Paul Krugman, "How to Destroy a Brand, Musk Style" *NY Times* (Dec. 30, 2022) at: www.nytimes.com/2022/12/30/opinion/elon-musk-tesla-democrats.html

116 Mark Harris, "Should Your Next Car Be Electric—or No Car at All? The Road to Decarbonization Hinges on Getting the Incentives Right" *Fixing Carbon: Anthropocene* (Feb. 16, 2023) at: www.anthropocenemagazine.org/2023/02/should-your-next-car-be-electric-or-no-car-at-all/ See also Angie Schmitt, "How to Go Car-Free—or Car-Light—in Middle America" *Vox* (Feb. 9, 2023) at: www.vox.com/future-perfect/23578481/how-to-live-without-a-car

117 *Washington Examiner*, "Yes, Environmentalists Want to Take Your Car Away—Just Slowly" (Feb. 11, 2023) at: www.washingtonexaminer.com/opinion/editorials/yes-environmentalists-want-to-take-your-car-away-just-slowly

118 See Joseph Stromberg, "The Real Reason American Public Transportation Is Such a Disaster" *Vox* (Aug. 10, 2015) at: www.vox.com/2015/8/10/9118199/public-transportation-subway-buses

119 Elizabeth Kolbert, "Why S.U.V.s Are Still a Huge Environmental Problem" The World Is Moving Toward Heavier Cars at a Time When It Should Be Doing Precisely the Reverse *The New Yorker* (Mar. 3, 2023) at: www.newyorker.com/news/daily-comment/why-suvs-are-still-a-huge-environmental-problem

120 A bold approach to promoting public transit has been launched in Germany. Germany has offered a so-called Deutschland-Ticket equal to $55 a month entitling a ticket holder to "unlimited travel on all city buses, subways and trams in every municipality across Germany." The goal is "to reduce carbon emissions by making it easier and cheaper for people to use public transport." *Carbon Almanac Network* (Aug. 18, 2023). See "Germany's Cheap Transit Ticket Is Boosting Train Trips, DPA Says" at: www.bnnbloomberg.ca/germany-s-cheap-transit-ticket-is-boosting-train-trips-dpa-says-1.1949524

121 David Zipper, "Electric Vehicles Are Bringing Out the Worst in Us: The Downside of Heavy, Overpowered Trucks and SUVs" *Atlantic Monthly* (Jan. 4, 2023) at: www.theatlantic.com/ideas/archive/2023/01/electric-vehicles-suv-battery-climate-safety/672576/

122 David Wallace-Wells, "Electric Vehicles Keep Defying Almost Everyone's Predictions" *New York Times* (Jan. 11, 2023) at: www.nytimes.com/2023/01/11/opinion/electric-vehicles-sales-growth-tesla.htm

123 Ted Nordhaus and Ashley Nines, "Don't Expect Mass Adoption of Electric Cars Anytime Soon" *Wall Street Journal* (Apr. 13, 2023) at: www.wsj.com/articles/dont-expect-mass-adoption-of-electric-cars-anytime-soon-fc60d894

124 "15 Questions with Lester Brown" *Harvard Crimson* (Nov. 14, 2013) at: www.thecrimson.com/article/2013/11/14/15q-lester-brown/

9   Investment for a Green Retirement

Investment for a green retirement means eco-investing or investing in companies that are environmentally sensitive. It can include investing in green mutual funds or other ways of holding stocks or bonds for environmentally friendly companies. It also includes banking and holding pension funds accountable for green investment policies. In this chapter we address all these issues.[1] What is clear now is that climate is a retirement issue. The challenge is how to turn worry into action.[2] A green retirement can include active steps on climate advocacy and sustainable living, but it also includes how we save and invest our money.[3] The bottom line is: money matters and we want our investments to follow our values.

Here's a question: Do you think of yourself as an investor? Maybe not. If you're *not* part of the top 1% in wealth, then you're among the 99%. In that case, you may not think much about investments at all. Even worse, up to half of Baby Boomers nearing retirement have no savings for retirement at all. If you do have some money, then some of it is probably in a bank. That makes you an investor, as we will see. You're either retired now or thinking that someday you could be retired. If you have money and if you want to retire, then you need to be thinking about a green retirement because climate change will shape the world in which retirement takes place. And when it comes to money and investment, you're either part of the problem or you're part of the solution.[4]

People over age 60 own the vast majority of stocks, either directly or through mutual funds. But, when trying about avoid investment in fossil fuels, how do we make investments reflect our values? How do we make our money matter? We can protest to big investment platforms, like Vanguard or Fidelity. Protests help gain their attention to do the right thing. But, right now, how do we act to fix our funds or ask our investment advisers to do this for us? Even if we have an investment advisor or professional expertise, we need help most of all just to ask the right questions. There is also guidance from organizations offering objective evaluation for individual companies and investment platforms that can make our money count. In this chapter, we will explore all these elements of what it means to make individual choices for investment for a green retirement, at whatever age we are.

DOI: 10.4324/9781003345992-13

In this chapter we look at the various ways in which investment for retirement takes place and then ask the question: How do these investments affect climate change? Could these investments be different to promote a green retirement? We will look at banks and financial Institutions such as credit unions or insurance companies. We will consider pension funds that provide a stream of retirement income. We will look at stocks and bonds and at mutual funds as vehicles for investment. And finally we will consider green-washing and alternatives to it: that is, investments claimed to be "green" or "sustainable" and environmentally appropriate places for your money.

We can approach the money-and-climate challenge in two very different ways. To "do-the-right-thing" can mean avoiding choices that create climate pollution. This approach of prudence makes a lot of sense when it comes to climate change. It reminds us that investors may be underestimating the risks of climate change.[5]

Prudence is one way to approach investment for a green retirement. The other is the opposite: to think about how climate change can open up unprecedented opportunities for investment. BlackRock is the world's largest asset manager, with $10 trillion currently under management. The CEO of BlackRock, Larry Fink, said this about climate change: "I believe the decarbonizing of the global economy is going to create the greatest investment opportunity of our lifetime. It will also leave behind the companies that don't adapt, regardless of what industry they are in."[6] As we've seen, industry is essential for the climate transition and the path to net-zero emissions runs through industry, so investment is essential here.[7]

The most important to know about investment for retirement is that most people don't do it at all. They just don't save for retirement and those who do save actually don't save much.[8] If they're lucky, they're enrolled in a pension plan of some kind, so they don't have to make decisions. Most importantly, they don't have to decide whether to spend or save, which is why pension plans, public or private, are a good thing. Investment depends on saving and savings do happen, just not as often as people believe it should. At least when they reach retirement age, people often say they wish they'd saved more.

There are solutions to this problem. There are apps like Acorns which will round up your spare change to the nearest dollar and then auto-invest it: for example, if you buy a coffee for $4.25, they'll round the purchase up to $5 and invest the $0.75 for you. But even saving for retirement doesn't mean investing the money wisely. When it comes to the stock market we are all subject to crowd psychology. In theory, wise investment should mean to buy low and sell high, but people often do exactly the opposite. Financial observers will tell you that Wall Street knows only two emotions: fear and greed. But we get the timing backward, which is what crowd psychology does.

Again, there are solutions: for example, dollar-cost-averaging, which forces us to invest the same amount each month, thereby overcoming crowd psychology. But where to invest? Warren Buffett, who knows a thing or two about investment, tells people just to buy index funds: aim for average gains in the stock market, which, over time, can be impressive. Wise financial advisers will tell you: it's not market timing, but time in the market that brings success.

What about climate change? How can we align our values with our investments? That's what we need to look at now because that's how we exercise individual agency—how we make a difference—when it comes to climate change.

Banks[9]

If you have a checking or savings account in a bank, then you're already an investor, probably in the fossil fuel economy, even if you don't know it. Your investments involve fossil fuel companies with an impact on climate change. In fact, banks are among the world's largest fossil fuel lenders.[10] Do you say you don't want to be an investor in fossil fuel companies? Maybe it's time for a different bank.[11] Third Act and other climate advocates have urged exactly that course of action. They've found a response among regulators who are beginning to tell fossil fuel companies to prepare for extreme weather events and anticipate a clean energy transition. For instance, financial regulators in New York State have told banks and mortgage companies to assess and manage threats they face from climate change.[12] While waiting for steps by policymakers and regulators, there are options for individual action. Some believe that so-called Green Banks are the answer.[13]

The big American banks—JPMorgan Chase, Citicorp, Bank of America, and Wells Fargo—are financing the fossil fuel industry, and they're doing it on a bigger scale than any other banks in the world. These banks continue to do so even in the face of climate catastrophe. For this reason Third Act and other climate advocacy groups have urged people to protest and to move money to other financial institutions to promote renewable energy.

According to Al Gore's Climate Reality Project, the world's largest banks have continued to support funding for fossil companies, pouring nearly $670 billion into fossil fuel companies in the past year. We can always shift money out of the big banks (Chase-Morgan, Bank of America, Citicorp, Wells Fargo). There is new pressure to do that: for example, the Fossil Free Finance Act (FFFA), introduced in Congress, would make the Federal Reserve force big banks to stop financing greenhouse gas emissions by 2050.

Why such emphasis on banks? For the simple reason that most people do have bank accounts. Not everybody owns stocks or mutual funds, but people typically have bank accounts and may also have accounts with insurance companies. All these financial institutions can be pressured to recognize the climate threat and to avoid underwriting fossil fuel projects. Groups like the "Stop the Money" Pipeline have identified ways of putting pressure on a viable climate future. Whether investing in a green retirement or becoming a climate-conscious consumer, there are things we can do as individuals.

In 2023 Third Act launched a massive consumer campaign to persuade the biggest banks to stop lending money for fossil fuel extraction. This initiative against the "dirty banks" reached a high point on March 21, 2023, when over a hundred Third Act demonstrations took place across the continent in 30 states and the District of Columbia.[14]

The big banks have provided more than a trillion dollars in loans to the fossil fuel industry and environmental groups emphasize that big banks are among the world's largest fossil fuel lenders.[15] For that reason Third Act has organized protestors to tear up their credit cards in order to send a message to the banks that funding the fossil fuel industry needs to stop. The biggest banks targeted were Chase-Morgan, Citicorp, Bank of America, and Wells Fargo. Third Act has also compiled a list of Responsible Finance Resources and blogs in order to help move money away from these big banks.[16] In 2023 Congress began consideration of the FFFA, which would demand that the Federal Reserve by 2050 force the big banks to stop their financing of greenhouse gas emissions.

Pensions

Pensions are also a channel for financing fossil fuel producers and other producers of greenhouse gases. Pensions for retirement are still widespread even if not universal. A few pensions came into being in the 19th century, but it was not until the mid-20th century that guaranteed pension plans became popular in both government and corporations. They are described as "defined-benefit plans" because they promise an income, generally for life with assets provided by the employer. In these plans employees get a promise of a fixed income but have little say in how the employer makes investment decisions.

In the last decades of the 20th century a different form of pension plan became more popular: "defined-contribution plans," which employees or individuals contribute their own money. 401(k) plans and similar vehicles offered people the option of shielding their current income in favor of retirement. Under defined-contribution plans people can have more freedom to decide on where the money is invested. The result is a greater opportunity to choose climate-sustainable investment vehicles.

As they gained acceptance, 401(k)s and other defined-contribution options became popular both with big private-sector employers and with individuals with enough income to save for retirement. For both defined-contribution pension plans and annuities, individuals can gain more control over their investments.[17] In these cases, it has become possible for individuals, retired or preparing for retirement, to find sustainable or green investments for retirement income. However, doing so requires considerable knowledge and skill to make the right choice.

When we think about pensions and sustainable investment, it becomes clear that individual pension holders have little choice. As individuals they have little influence on investment policy and, unlike holding a bank account, they have little choice about leaving a pension fund for another one. For this reason, there is interest in pushing pension funds, public or private, toward divestment from fossil fuel companies.

The situation for pension holders is comparable to institutions that hold large endowment assets invested in fossil fuel companies. For example, Harvard University reported that in 2022 climate solutions investments would account for 1% of its total endowment, roughly doubling what it reported for the previous year.[18] It is expected to increase this level in the future. Harvard spoke favorably about directing more investments to support the transition to a green economy, including investments in technologies such as carbon capture and sequestration. Eventually, pressure forced Harvard to act on divestment.

Harvard's decision to withdraw from fossil fuel investment reflected changing financial realities. But it also reflected public pressure and is reminiscent of pressure to boycott investment in South Africa at a time when Apartheid was in place. One of the strongest public opponents of Apartheid, Bishop Desmond Tutu, urged a boycott of fossil fuel companies, as reported by Bill McKibben when he founded the Third Act group on climate advocacy. In many respects, the push for divestment by pension funds echoes a similar effort made decades ago during the time of Apartheid in South Africa, when there was pressure on financial institutions to divest from enterprises in South Africa.

A released report from the University of Waterloo, in Ontario, Canada, in partnership with Stand.earth found that the California state pension fund, CalPERS, managed to lose $4.7 billion over the past decade, or $3,163 per pensioner, by staying invested in fossil fuel and that the smaller CalSTRS managed to lose $4.9 billion, or $5,114 per beneficiary.[19]

Some pension funds don't have to wait for stranded assets to lose money in a time of climate change. Some efforts are directed not for investors but for fossil fuel companies themselves. ExxonMobil is the largest Western oil company, but, compared to other oil companies, it has not been a leader in taking on climate issues. But in May 2021, a shareholder revolt resulted in three new directors more supportive of climate issues on the ExxonMobil Board. By January 2022, the CEO of ExxonMobil, Darren Woods, announced that the company was committed to a lower-emissions future.[20] The ExxonMobil website now proclaims that the company supports reducing carbon dioxide emissions and supports work on alternative fuels, carbon capture and storage, and reducing methane.

So-called Climate Finance is an area of growing interest to investors. Many financial institutions have pledged to eliminate deforestation from investment portfolios since forests are an important element in removing carbon dioxide from the atmosphere. One example of this is the Glasgow Financial Alliance for Net Zero, whose name was linked to the 2021 Conference on Climate Change held in that city in Scotland. The Financial Alliance is co-chaired by a former governor of the Bank of England, and it includes financial institutions that hold a staggering amount of assets: $130 trillion, according to the best estimates. The goal of this coalition is to reduce emissions by cutting back on lending and investing in companies that damage the climate.[21]

9.1   Yvon Chouinard: Eco-Elder

Yvon Chouinard founded Patagonia, the world-famous outdoor clothing and gear company, which began in 1973 and became a $3 billion company. In 2022, Chouinard, along with his wife and children, turned over their entire company ownership to a nonprofit trust, the Holdfast Collective, whose supreme mission is to work against climate change.[22]

Chouinard, at age 85, remains a billionaire. But his purpose in old age is to save planet Earth. In his own words:

> Everything man does creates more harm than good. We have to accept that fact and not delude ourselves into thinking something is sustainable. Then you can try to achieve a situation where you're causing the least amount of harm possible.[23]

Chouinard wants other business leaders to see that doing the right thing can end up making more money. His motivation is "just believing in karma. It comes back every single time." In opposition to the mindless growth of capitalism, he wants a private enterprise that respects the environment: for example, through regenerative agriculture and other steps that reflect long-term thinking.

In the last years of his life, Chouinard is the kind of Eco-Elder who will never retire for making the world better than it has been.

Stocks and Bonds

Another approach to investment for retirement is to buy directly individual stocks or bonds, generally through a broker or financial advisor. This approach has the advantage of giving great individuals more choices to find sustainable or green investments. But, as with defined-contribution investment plans, considerable knowledge and skill are required to be sure that the investments are actually sustainable. We will return to this point further and consider the challenge of "greenwashing" where investors are given a mistaken impression that an investment is actually favorable for climate protection.

Mutual Funds

Mutual funds have the great advantage of wide diversification of stocks or bonds along with low cost, which is why they are attractive. Mutual funds may be active, in which fund managers make decisions about buying individual stocks or bonds. Or funds may be index funds, in which managers simply buy a broad cross-section of stocks. Renowned investor Warren Buffet has said that he would recommend people wanting to have a productive investment in the stock market to buy an index fund: that is, a mutual fund consisting of all stocks within some broad category, such as the Standard & Poor's (S&P) 500 or the Dow Jones index. Index fund investing, over a long period, for a buy-and-hold investor would enable one to match the growth of the stock market overall. For example, the S&P 500 return since 1980, more than 40 years, has come to over 11% per year. Even after accounting for inflation, the return is more than 8% per year. Index fund investment appeared in the 1970s under the leadership of the Vanguard family of funds.

At this point, we have to recognize that index fund investment, however attractive it seems, is not an investment for a green retirement. It cannot be for the simple reason that an index contains fossil fuel companies, like Chevron and Shell, and also includes banks, like Citicorp or Wells Fargo, that are investors in fossil fuel producers.

Index fund investment, like buying other mutual funds, typically involves investment through big asset managers, such as Vanguard, Fidelity, or Black Rock. Many climate advocates are unhappy with these big asset managers because most of the mutual funds, including index funds, are invested in fossil fuel producers. What is an individual investor to do about that? One option is simply to avoid big asset managers and select individual stocks and bonds that are climate-sustainable. Buying individual investments means getting away from big asset managers. Another option is to identify climate-sustainable mutual funds offered by big asset managers, if that choice is available. In this approach, investors for a green retirement would not necessarily leave big asset managers. Instead, investment for a green retirement would mean shifting assets more and more into climate-sustainable mutual funds offered by big asset managers.

Investment for a green retirement is not simple and it's easy to be fooled. Remember Bernie Madoff? I'm a lifelong New Yorker and got quite familiar with streetside con games. I never invested with Bernie Madoff and therefore never lost money. But think of Madoff when you hear about appealing ideas like carbon offsets as a way of using financial comparisons to compensate for products that pollute the earth's atmosphere. Carbon offsets have been hailed as a fix for climate change—but the market's largest trading firm sold millions of credits for carbon reductions that weren't real. Watch before they take your money.[24]

When it comes to investment for a green retirement there are parallels with other choices in the marketplace: for example, buying food. For example, if Safeway offers an organic food section, should I continue to shop at Safeway even though most of their groceries are NOT organic? That's an analogy for Vanguard, Fidelity, and Black Rock. The problem is that it's not always easy to identify "green" mutual funds. Should we continue to invest even if most of their offerings are not what we want and perhaps illustrate greenwashing?

This is a question investors need to answer. Does "piling on the pressure" for big asset managers means abandoning these asset managers entirely? That approach is what the Third Act, for example, recommends in dealing with big banks like Citicorp, Wells Fargo, Chase-Morgan, and Bank of America. By leaving these big banks there's another option, for instance, of going to a credit union that has no fossil fuel investments at all. The opposing argument is to stay with one of the big banks in order to "pile on the pressure" because existing customers will have more access.

Greenwashing

In thinking about investment for a green retirement, we need to confront a big problem: namely, knowing whether a company or a mutual fund is "green" or "eco-friendly" despite what they say about themselves.

"Greenwashing" describes what a business does when it claims that its product or service is environmentally positive: for example, that it is working against harmful climate change. For example, a hotel might claim to encourage guests to reduce the use of towels to "save the environment," when the actual motive is for the hotel to save on laundry costs. But maybe reducing energy use for laundry actually does have a climate benefit. What about the case of a fossil-fuel company, like Chevron, that claims to be eco-friendly when its products are the main driver of global warming?

Greenwashing has been around since the 1960s. But as the climate crisis has attracted more attention, more and more companies are trying to rebrand or repackage their products to appear environmentally positive. How can we tell what companies are telling us?

The most appropriate response to greenwashing was what President Ronald Reagan said about arms control: trust, but verify. And the familiar saying "Actions speak louder than words" is always appropriate. Consumers and investors need help to find out the truth in the marketplace.

The Federal Trade Commission (FTC) has offered guidelines meant to protect consumers from greenwashing:[25]

- Packaging and advertising should explain the product's green claims in plain language and readable type in close proximity to the claim.
- An environmental marketing claim should specify whether it refers to the product, the packaging, or just a portion of the product or package.
- A product's marketing claim should not overstate, directly or by implication, an environmental attribute or benefit.
- If a product claims a benefit compared to the competition, the claim should be substantiated.

The bottom line here is that, whether you're a consumer or an investor, you're on your own. You can be misled and you need more than just the FTC's guidelines to verify whether a product is worthy of your trust.

Climate activists have been vocal in criticizing greenwashing not only by private companies but by governments engaged in talks to stop climate change. Greta Thunberg[26] said that such climate talks are becoming nothing more than a "Greenwash Campaign": "We've been greenwashed out of our senses. It's time to stand our ground".[27]

When it comes to investment, so-called green bonds are a good example of the challenge. Green bonds refer to money going into investment that claims to help save the plane. By some estimates, more than a trillion dollars has been invested in green bonds in recent years.[28]

More and more companies are getting the message from consumers that they should do something about climate change. 80% of corporations examined in a recent study of greenhouse gas emissions turn out to have a formal board committee devoted to climate sustainability committee. There is evidently a belief that such action could increase a company's market value. But it's also possible that companies that ignore climate risk will lose market value.[29]

It turns out that companies that climate change initiatives have a higher value in the stock market. But, ironically, also tend to have higher levels of greenhouse gas emissions.[30] Trust but verify turns out to be very good advice, indeed.[31]

Climate advocates have warned that fossil fuel companies are actually a dangerous investment choice for retirement because fossil fuels in the ground will become "stranded assets" and lose their value. Yet the response of fossil fuel companies has raised the peril of greenwashing:

> Environmentally-conscious investment products are incredibly popular, but such labels in finance are notoriously rife with greenwashing. The Network for Greening the Financial System, a coalition of more than 100 central banks and financial regulators, was emphatic in a 2019 report about the need for a global taxonomy to identify which activities would contribute to a low-carbon economy, and which activities were exposed to losses in a shift away from damaging fossil fuels or from the effects of climate change itself. That, the central bankers said, would be an essential building block for their work.[32]

Is it the case that there are a few "bad apples" while other companies are now trying to "do the right thing?" We can look at the big oil companies to find the answer. Over half of Big Oil's ads claim that they're on the side of clean energy and reducing emissions. That's the finding from the London-based group InfluenceMaps. The group looked at the big companies (e.g., Chevron BP and Shell) and found that in one year the companies spent $750 million went for greenwashing claims.[33] Companies don't want to be stigmatized and suffer "reputational damage" from climate-damaging activities. They certainly don't want to be found guilty of greenwashing. So they come up with sophisticated ways of shielding themselves from the market consequences of greenwashing.[34] But consumers and investors may not always punish climate polluters for their lies.

CDP is an international agency for assessing global corporate behavior around sustainability and the environment. They looked at companies around the world and found that almost all climate-related disclosures are inadequate. The results are stark: fewer than one in 200 companies have credible climate plans.[35]

Older Americans, as a group, control most of the wealth in the USA, which means that older people have a special challenge in determining whether their investments or their consumer purchases are going to products and services that damage the climate. Greenwashing, as some analysts have said, is becoming the new battleground for climate misinformation.[36] The end of greenwashing is not coming any time soon. It will continue to be a challenge for anyone who wants to invest in a green retirement.

ESG

"ESG" stands for environmental, social, and governance criteria that investors use to evaluate companies. During 2022, a year of decline in both stocks and bonds, ESG mutual funds displayed resilience in attracting new investors.[37] In the USA, as in Europe, sustainable funds attracted increasing numbers of investors. But ESG has become a political target from the Republican Party:

> GOP lawmakers claim that considering climate risks while making investments imposes "woke" values and limits investment returns. Yet anti-ESG laws passed in Kansas, Oklahoma, and Texas (in 2023) were estimated to have cost taxpayers up to hundreds of millions of dollars. That's partly because most Wall Street banks and businesses still employ ESG strategies.[38]

Can we count on ESG funds as a vehicle for climate advocacy? ESG funds, along with the broader stock market in the S&P 500, have declined in value at times. In contrast to fossil-fuel companies, which have been big winners, partly because of Russia's invasion of Ukraine and pressure on the international oil market. Investors in sustainable mutual funds may not be worried about short-term returns since they hope and believe that a transition away from fossil fuels will happen in years to come.

What about aging and ESG funds? Long-term investment is less relevant to older investors who have limited life expectancy. And here's another problem. Oil companies may have done well in the short term, but climate-sensitive investors may believe that this will change as net-zero deadlines come closer. According to data from Morningstar, the definitive rating source on mutual funds, green investment funds are likely to retain investors who are willing to stay with their commitment on behalf of climate sustainability. But commitment has limits, too. Older people, as a rule, have legitimate concerns about risks and the time horizon for investment success.

Another factor to consider is the pushback from political forces against ESG funds of all kinds. Republican political leaders in 2022 started a campaign against what they termed "woke capitalism," which includes ESG funds. By 2023, Republicans in Washington and state capitals around the country were up in arms against what they termed "Woke Capitalism." In many so-called red states, Republican politicians loudly complained about "ESG investing" and attacked Wall Street financial firms. Partly as a result Vanguard pulled out of its commitment to be involved with net-zero investment, while Black Rock was criticized by some state pension funds because of concern to take account of climate change in investment decisions. In short, conservative politicians have argued that ESG is diverting asset managers from their duties to investors, and may even amount to illegal collusion.[39]

The advocacy group Green America has argued that attacks on climate-related investing could be damaging. They make the point that such attacks on ESG criteria for investment are similar to accusations about voter fraud or critical race theory. They are part of a "culture war" strategy that could hurt climate progress. In 2022 more than 500 shareholder resolutions were filed on a range of ESG, and climate change is high on the list of public concerns.

One of the big campaigners against ESG was the Republican seeking the presidential nomination, Vivek Ramaswamy. Ramaswamy went so far as to insist that "climate change in a hoax." Why would a candidate say such a thing? As usual, the find answer follows the money:

> Ramaswamy has reported more than $50 million in holdings in [the investment fund] Strive, which to date, is mostly a fund to promote fossil fuel development . . . Ramaswamy pushes rejection of the "climate change agenda." He says it's because he's not "bought and paid for." The reality is the complete opposite.[40]

Finding Green Retirement Investment Options

In light of the challenges from greenwashing and claims about ESG mutual funds, how can someone preparing for retirement find a sustainable or green investment choice? The answer is it can be done but only with difficulty. Yet it is possible. Investment in a green retirement through sustainable mutual funds requires careful research. Even if one has a personal financial advisor, it requires knowing what are the right questions to ask and being willing to pose those questions with sound knowledge. For example, Vanguard already offers seven ESG funds that are environmentally friendly and mostly, if not entirely, fossil free.[41]

The respected financial research firm Morningstar has a valuable part of their website analyzing alternative investments as well as mutual funds and ETFs according to a sustainability standard. In this metric each type of investment is given a rating between one and five "planets," where five planets indicate the most favorable to ESG goals. The nonprofit advocacy firm As You Sow offers a website evaluating 3,000 different mutual funds and ETFs according to respect to environmental and social goals.

The Forum for Sustainable and Responsible Investment (U.S. SIF) hosts an online chart providing data on sustainable mutual funds offered by U.S. SIF member firms.[42]

Whether through divestment or through rigorous research to find sustainable investments, it is possible but not easy to invest in a green retirement. As we have seen in political debates around ESG, there is no way to separate politics and public policy from investment decisions. Individuals can try to act on their own, but eventually citizen climate action will be called for, and that is the subject of the following chapter.

9.2   Christiana Figueres: Eco-Elder

Figure 9.1 Christiana Figueres. Public domain.

Christiana Figueres is a Costa Rican diplomat best known for her leadership in forging the 2015 Paris Agreement on climate change, a historical achievement that emerged after many less successful international COP meetings. She came by her diplomatic skills from childhood, as the daughter of her father, three-time president of Costa Rica, and her mother who was also a diplomat. Her climate advocacy was the legacy of previous generations on behalf of generations to come.

Figueres was given the prestigious Joan Bavaria award by the non-profit group Ceres and Trillium Asset Management to honor her work as a global leader working to move capital markets toward a system that balances economic prosperity with social and environmental concerns. In her collaborative work, she helped to bring together governments, corporations, investors, communities of faith, and nonprofit groups to achieve the Paris Climate Change Agreement, signed by 175 countries.

In a world of foreign policy conflict that easily leads to discouragement, Figueres is known as a voice of hope and optimism. She founded the Global Optimism group and is co-author of *The Future We Choose: Surviving the Climate Crisis*. She also co-hosts a regular podcast, "Outrage and Optimism." That name in itself expresses the combination of moral commitment and hope for the future that is her watchword. Beyond her diplomatic expertise, she has displayed a willingness to admit her own mistakes. In 2023 she said, "I thought fossil fuel firms could change. I was wrong. Their unprecedented profits over the past year have shown their unwillingness to adapt."[43]

Born in 1956, Christiana Figueres is now in her seventh decade and her own positive aging shows that she continues to be a voice around climate change in an aging society. About investment for a green retirement, Figueres' life underscores the importance of collaborative diplomacy with investors and businesses when it comes to leaving a legacy for future generations.[44]

Notes

1 https://en.wikipedia.org/wiki/Eco-investing
2 Steve Vernon and H.R. Moody, "Climate Change Is a Retirement Issue—How to Turn Worry into Action" *Market Watch* (June 28, 2022) at: www.market-watch.com/story/climate-change-is-a-retirement-issue-how-to-turn-worry-into-action-11656090714
3 John Wask, "Want to Live Longer? Refine Retirement with Active Green Aging'" at: www.forbes.com/sites/johnwasik/2023/12/08/want-to-live-longer-refine-retirement-with-active-green-aging/
4 There are good sources for guidance on investment for a green retirement: https://investyourvalues.org/

   https://stopthemoneypipeline.com/customers/
   https://thirdact.org/our-work/banking-on-our-future/#actions
   https://fossilfreefunds.org/funds?pg=2&srt=ussif
   www.carboncollective.co/
   www.bankingonclimatechaos.org/
   https://bankonourfuture.org/
   www.workforclimate.org/post/the-financial-argument-for-a-climate-safe-retirement-fund
5 "How Investors Are Underpricing Climate Risks" *Financial Times* at: www.nakedcapitalism.com/2023/08/investors-start-thinking-about-underpricing-of-climate-risk.html
6 Larry Fink Statement at: https://finance.yahoo.com/news/blackrock-green-transition-larry-fink-165311244.html. But actions by Larry Fink and Black Rock speak louder than words: "BlackRock Promised to Be Climate Conscious. The Joke's on Us." See:

https://www.msn.com/en-us/money/markets/blackrock-promised-to-be-climate-conscious-the-joke-s-on-us/ar-AA1qAhQ6?ocid=msedgdhp&pc=U531&cvid=336ce4e8cc3c4f6b960b48e4902bdc99&ei=63

7 "The Path to Net-Zero Emissions Runs Through Industry" at: https://theconversation.com/the-path-to-net-zero-emissions-runs-through-industry-218773
See also *Net-Zero Industry Tracker: INSIGHT REPORT* (Nov. 2023) www3.weforum.org/docs/WEF_Net_Zero_Tracker_2023_REPORT.pdf

8 USAFacts, "Half of American Households Have No Retirement Savings" at: https://usafacts.org/data-projects/retirement-savings

9 In this discussion, we use the term "banks" but the points made apply as well to credit unions and insurance companies where people have reliable access to cash.

10 "Banking on Chaos: Fossil Fuel Report 2022" at: www.bankingonclimatechaos.org/wp-content/themes/bocc-2021/inc/bcc-data-2022/BOCC_2022_vSPREAD.pdf

11 "How to Switch to Better Banks and Credit Cards: FAQs" at: https://thirdact.org/resources/how-to-switch-to-better-banks-credit-cards-faqs

12 https://subscriber.politicopro.com/article/eenews/2023/01/04/new-york-tells-its-banks-to-prepare-for-climate-change-00076246

13 " 'Transformational': Could America's New Green Bank Be a Climate Gamechanger?" *The Guardian* (Sept. 11, 2022) at: www.theguardian.com/us-news/2022/sep/11/green-bank-clean-energy-climate-change?CMP=Share_iOSApp_Other

14 "Older Americans Protest Against 'Dirty Banks' Funding Oil and Gas Projects" *The Guardian* (Mar. 21, 2023) at: www.theguardian.com/us-news/2023/mar/21/dirty-banks-protest-washington-chase-citibank-bank-of-america-wells-fargo   See also: "A 'Rocking Chair Rebellion': Seniors Call on Banks to Dump Big Oil" *N.Y. Times* at: www.nytimes.com/2023/03/21/climate/climate-change-protests-oil-banks.html?smid=nytcore-ios-share&referringSource=articleShare

15 "Banking on Chaos: Fossil Fuel Report 2022" at: www.bankingonclimatechaos.org/wp-content/themes/bocc-2021/inc/bcc-data-2022/BOCC_2022_vSPREAD.pdf

16 "How to Switch to Better Banks and Credit Cards: FAQs" at: https://thirdact.org/resources/how-to-switch-to-better-banks-credit-cards-faqs

17 Melissa Phipps, "The History of Pension Plans in the U.S." *The Balance* at: www.thebalancemoney.com/the-history-of-the-pension-plan-2894374 See also William Graebner, *A History of Retirement* (Yale University Press, 1980).

18 "Climate Solutions at 1% and Growing for Harvard" by Hazel Bradford, *Pensions and Investments* (Feb. 4, 2022) at: www.pionline.com/endowments-and-foundations/climate-solutions-1-and-growing-harvard

19 Bill McKibben, "California's Pension Funds Are Wrecking the Planet and Losing Billions. It's Quite a Trick" at: www.msn.com/en-us/money/markets/opinion-californias-pension-funds-are-wrecking-the-planet-and-losing-billions-it-s-quite-a-trick/ar-AA1d9lLF

20 *The Economist* (Feb. 12, 2022).

21 "Marketplace" *The Economist* (Nov. 11, 2021).

22 Sam Silverman, "Who Is Patagonia Founder Yvon Chouinard? Meet the Entrepreneur Who's Giving Away His $3 Billion Brand" (Sept. 15, 2022) at: www.entrepreneur.com/living/who-is-patagonia-founder-yvon-chouinard-net-worth-and-more/435446. See also David Gelles, "Billionaire No More: Patagonia Founder Gives Away the Company" *The New York Times* (Sept. 14, 2022), For more on the Patagonia story see David Gelles and Kenneth P. Vogel, "Patagonia's Profits Are Funding Conservation—and Politics" *N.Y. Times* (Jan. 30, 2024) at: www.nytimes.com/2024/01/30/climate/patagonia-holdfast-philanthropy.html

23 Jeff Beer, "Patagonia Founder Yvon Chouinard Talks about the Sustainability Myth, the Problem with Amazon—and Why It's Not Too Late to Save the Planet"

Fast Company (Oct. 16, 2019) at: www.fastcompany.com/90411397/exclusive-patagonia-founder-yvon-chouinard-talks-about-the-sustainability-myth-the-problem-with-amazon-and-why-its-not-too-late-to-save-the-planet

24 "The Great Cash-for-Carbon Hustle" at: www.newyorker.com/magazine/2023/10/23/the-great-cash-for-carbon-hustle

25 Louis Marmon, "What Is Greenwashing?" *Advance ESG* (Jan. 20, 2022) at: www.advanceesg.org/what-is-greenwashing/ See also Ivan Couronne, "New Carbon Accounting Rules Target 'Greenwashing'" *Phys Org* (June 26, 2023) at: https://phys.org/news/2023-06-carbon-accounting-greenwashing.html

26 Jenny Gross, "Greta Thunberg Says Climate Talks Are Becoming a 'Greenwash Campaign'" *New York Times* (Nov. 23, 2021) at: www.nytimes.com/2021/11/04/climate/greta-thunberg-cop26.html

27 Greta Thunberg, "Greta Thunberg on the Climate Delusion: 'We've Been Greenwashed Out of Our Senses. It's Time to Stand Our Ground'" *The Guardian* (Oct. 8, 2022) at: www.theguardian.com/environment/2022/oct/08/greta-thunberg-climate-delusion-greenwashed-out-of-our-senses

28 Felix Salmon, "The Trillion-Dollar Greenwashing Market" *Axios Markets* (Feb. 11, 2023) at: www.axios.com/2023/02/11/the-trillion-dollar-greenwashing-market

29 Sarah DeWeerdt, "Companies That Ignore Climate Change Risk Lose Market Value" *Anthropocene* (Apr. 23, 2024) at: www.anthropocenemagazine.org/2024/04/companies-that-ignore-climate-change-risks-lose-market-value/

30 Sarah DeWeerdt, "Can Corporate Greenwashing Be Proven Empirically? Maybe" *Anthropocene* (Feb. 28, 2023) at: www.anthropocenemagazine.org/2023/02/can-corporate-greenwashing-be-proven-empirically-maybe/

31 "What Banks Really Mean When They Put Trillions Into ESG" (Bloomberg) at: https://finance.yahoo.com/news/banks-really-mean-put-trillions-050001889.html

32 Kate Mackenzie, "Stranded Assets" *Bloomberg Green Daily* (Feb. 18, 2022).

33 InfluenceMap, "Big Oil's Real Agenda on Climate Change 2022" at: https://influencemap.org/report/Big-Oil-s-Agenda-on-Climate-Change-2022-19585

34 George I. Kassinis, et al., "Stigma as Moral Insurance: How Stigma Buffers Firms from the Market Consequences of Greenwashing" *Journal of Management Studies* (Sept. 27, 2022) at: https://onlinelibrary.wiley.com/doi/10.1111/joms.12873

35 Simon Jessup, et al., "Almost All Climate-Related Corporate Disclosures Are Inadequate, CDP Says" *CDP* (Mar. 3, 2022) at: www.reuters.com/business/sustainable-business/almost-all-climate-related-corporate-disclosures-are-inadequate-cdp-says-2022-03-03/

36 Roland Lloyd Perry, "'Greenwashing': New Climate Misinformation Battleground" *AFP Fact Check* (June 27, 2022) at: https://factcheck.afp.com/doc.afp.com.32AD4DZ

37 Buttonwood, "The Tenacity of ESG" *The Economist* (Nov. 19–25, 2022), p. 70.

38 "Republicans Ramp Up Their War on 'Woke' ESG Investing" From: "24 Predictions for 2024" *Grist* (Jan. 3, 2024) at: https://grist.org/culture/24-predictions-for-2024-climate-trends/

39 Marianne La Velle, "Republicans Are Primed to Take on 'Woke Capitalism' in 2023, with Climate Disclosure Rules for Corporations in Their Sights" *Inside Climate News* (Jan. 3, 2023).

40 Emily Atkin, "How Vivek Ramaswamy Makes Money from Climate Denial" *Heated* (Aug. 28, 2023) at: https://heated.world/p/how-vivek-ramaswamy-makes-money-from

41 For details, see: https://investor.vanguard.com/investment-products/esg

42 For Mornningstar, see www.morningstar.com/try/esg-data For As You Sow, see: www.asyousow.org/ For The Forum for Sustainable and Responsible Investment (US SIF) see: www.ussif.org/ Recommendations here are drawn from Steve Vernon, "How Can Retirees Invest Sustainably?" *Forbes* (Feb. 3, 2023) at: www.forbes.com/sites/stevevernon/2023/02/03/how-can-retirees-invest-sustainably/?sh=757d655074ba

43 Christiana Figueres, "I Thought Fossil Fuel Firms Could Change. I Was Wrong" *Aljazeera* (July 6, 2023) at: www.aljazeera.com/opinions/2023/7/6/i-thought-fossil-fuel-firms-could-change-i-was-wrong

44 "Turning from Peril to Possibility: Ecological Superhero Christiana Figueres on the Spirituality of Regeneration" at: www.themarginalian.org/2023/11/12/christina-figueres-on-being/

"Christiana Figueres: Ecological Hope, and Spiritual Evolution"

(On Being with Krista Tippett) at: https://onbeing.org/programs/christiana-figueres-ecological-hope-and-spiritual-evolution/

10 Citizen Climate Action

"Do what you can, with what you have, where you are."
—Theodore Roosevelt

In many ways, the battle for public opinion on climate change has already been won: Americans understand that climate change is a reality and that human beings are the cause.[1] From lived experience, Americans have already come to see that the Four Horsemen of the Climate Apocalypse are already here. Even long-standing proponents of denial, like fossil fuel companies, have shifted their rhetoric from outright denial to delay and evasion.

But if Americans understand the basic challenge, there remains a stumbling block. It is the question, Will I make a difference? The public opinion surveys show that three in five Americans care about saving the environment as much as they care about saving money. But what will they do about it? Part of the problem is what I have called "learned helplessness:[2]" namely, a belief that "There's nothing I can do" or "What can one person do?" Those statements are a prescription for complacency.

I have friends who are so distressed that they have gone on a "news fast." Whether the issue is climate or Ukraine or the Middle East: it's all too much, they say. For climate advocacy, the great enemy is complacency and pessimism: "Nothing works. Why bother?"

How we think about politics is the problem here, as Robert Reich, former Secretary of Labor, aptly put it. One person interviewed by the *Washington Post* put it this way: "I can't really speak to anything [Biden] has done because I've tuned it out, like a lot of people have. We're so tired of the us-against-them politics."

Focusing on government dysfunction ignores what actually does work (e.g., the Environmental Protection Act). Those of us who are elders remember what it was like before Earth Day (1970) when smog engulfed Los Angeles and when the river in Cleveland caught fire. Ignoring slow but steady results makes it more likely for America to fall into neofascism and authoritarianism. The advice of Theodore Roosevelt remains the best answer to feelings of despair and learned helplessness: Do what you can, with what you have, where you are.

DOI: 10.4324/9781003345992-14

In too many cases, people who recognize the problem put their attention on individual actions, a point discussed earlier in the chapter on becoming a climate-conscious consumer. All good, all necessary, but not sufficient. David Wallace-Wells, author of the devastating book *The Uninhabitable Earth: Life After Warming*, put it well. "The climate calculus is such that individual lifestyle choices do not add up to much, unless they are scaled by politics."

The challenge, then, is how to help people move from learned helplessness—the politics of complacency and passivity—into a politics that has results, a "scaled up" politics that is sustainable beyond short-term gains and losses. This strategy is what Terry Eagleton summed up in his book *Hope Without Optimism*. As Greta Thunberg said "Once we start to act, hope is everywhere. So instead of looking for hope, look for action. Then, and only then, hope will come."

The challenge is to find what constitutes the right action or to find what Buddhists call "skillful means" (*upaya-kausalya*) which means how an enlightened person tailors a message for a specific audience. In this chapter we look at a series of recent efforts for climate advocacy, some by elders, others by young people, and others by groups that cross generational lines. We will look at successes and also at failures and at mixed outcomes.

An important question for skillful means is the question of time: we have limited time to act to prevent some of the worst possible outcomes of global warming and environmental devastation. Many bad outcomes have already occurred; others, like rising sea levels, will come about even if we are successful beyond our wildest dreams: even, if, for example, all fossil fuel use vanished at midnight tonight. The question of time involves elders who act for future generations, hopefully to act in a spirit of generativity, or what the Greek proverb summed up this way: "A society grows great when those who are old plant trees whose shade they know they shall never sit in."

Many have heard the famous statement by Margaret Mead: "Never doubt that a small group of thoughtful, committed citizens can change the world; indeed, it is the only thing that ever has." But do all of us really believe what Margaret Mead said? Do we act on it and take seriously the importance of building strong personal ties with other activists on climate change? If not, then we're in trouble. Why? Because it's hard to sustain motivation for a cause that seems as diffuse and all-encompassing as climate change. Firefighters, and even strangers, risk their lives to save people from a burning building. But to save "possible people" from a catastrophe that is years away—and decades in the making? Not likely. Becoming a Good Ancestor is harder than one might imagine.

Edmund Burke long ago referred to the "little platoons" or small associations that we belong to in life, beginning with the family but also including many "mediating structures" in the neighborhood, the workplace, faith communities, and so on.[3] Without the strength of "little platoons" soldiers in actual platoons would not risk their lives in battle. They do it for people they know and care about. Large causes, like climate change, are important. But changing the world, as Margaret Mead calls for, will demand something else, and means personal ties for people who care as we do.

We also need to realize that change begins at many different levels. The environmentalist slogan "Think globally, act locally" recognizes this fact. We forget it at our peril. Very few of us who are climate advocates occupy positions of power or command in business or politics. But we can have influence in these areas, which is why it is important to be a climate-conscious consumer and why investment in a green retirement makes sense.

Individual acts in the marketplace or investments in themselves will not change the world. So it is natural that people fall victim to the doubtful voice, "What can one person do?" Political action, citizen action for the climate crisis, is what is called for. But citizen action requires that each of us will "walk the talk" and sustain ties—in those "little platoons"—with others who are walking the same path. This common action is what makes hope something more than sentiment.

In this chapter we will look closely at those platoons, both little and larger, that have mobilized for the battle around climate change. These include aging organizations, such as Third Act and Elders Climate Action (ECA), as well as AARP, the largest of all aging organizations, with 38 million members. We will also look at groups of younger people, such as Citizens Climate Lobby, and at youth-based initiatives such as Fridays for the Future and the Sunrise movement.

It is a paradox, but an essential point, that in an aging society, young people can be the vanguard of citizen action on climate change, even when they're not even old enough to vote.[4] Many young people, facing a planetary crisis, feel that action is now up to them. In August 2023, a Montana state court decided in favor of young people who filed suit claiming that the state by promoting fossil fuels had violated their right to a "clean and healthful environment." The plaintiffs filed this case when they were between ages 2 and 18. One of the plaintiffs was Badge Buss, age 15 in 2023. He said "The fact that kids are taking this action is incredible. But it's sad that it had to come to us. We're the last resort."

Badge Buss's statement was reminiscent of Jerry Garcia of the Grateful Dead who was quoted as saying, "Somebody needs to do something. It's just incredibly pathetic that it has to be us." "That mix of pride and exasperation [from the plaintiffs] is not uncommon among young climate activists. Many are energized by what they see as the fight of their lives, but also resentful that adults haven't seriously confronted a problem that has been well understood for decades now."

"It seems to me that adults just get so cynical and lose their imagination" is a statement by Tia Hatton, who was 18 when she joined the Montana lawsuit that resulted in a judicial decision against climate change.[5]

"The Montana climate kids' lawsuit has energized activists, including this one".

The challenge for intergenerational collaboration in climate advocacy remains uppermost,[6] as I argued in the chapters on "Who's to Blame for

Climate Change?" and "What Should We Do?" Karl Pillemer, founder of the Cornell Clearinghouse on Aging and Climate Change, put the more concisely when he said that a society in which members of different generations do not interact is a dangerous experiment. Are we embarking on that dangerous experiment today? A better role for older people is based on taking responsibility. One elder climate activist put it this way:

> This is my time. It's my generation that made the mess . . . We want to take care of our children and our grandchildren but we did such a poor job of not cleaning up the mess . . . I can do it now. There's nothing to stop me.[7]

Intergenerational collaboration is essential. But we must also consider groups like Extinction Rebellion, which are not organized around age but around a more radical agenda, much in the tradition of Green Peace and other environmental advocacy groups.[8] Success in citizens' climate advocacy will depend on multiple tactics to achieve the goal, even if not all at once. Not all will act the same way Extinction Rebellion has done, but multiple tactics are necessary.

The advocacy approach for citizen climate action was stated by Vice President Kamala Harris: "Talk with your friends, your neighbors, your colleagues and make sure they understand the real consequences . . . Let's organize, activate our communities and remind folks of what's at stake." Harris was speaking about the federal budget default crisis of 2023, but the message is the same for climate advocacy.[9]

This approach is exactly what ECA and Third Act have been doing in recent years and what they continue to do. Ezra Levin, co-founder of the anti-Trump group known as Indivisible, has characterized their successful political strategy as one involving outside grassroots pressure to open up space for advocacy in negotiations outside of public view. When successful, this strategy forces concessions to pressure, even where some compromises have to be made. But this "Inside-Outside" strategy only works when local advocates work together with national leaders in effective campaign teams. In such cases, outside pressure can make a big difference.

When Franklin Roosevelt was pushing forward the New Deal, he met with activists and advocates who were urging him to a specific course of action, more radical than what anyone was prepared for. FDR's reply was simple: Get out there and organize people and have them make me do the things you want me to do. Instead of passivity or complaining, FDR was urging activism that would push political leaders to make the policy changes they wanted. The situation is no different for climate change in the current decade. An Inside-Outside strategy that requires both national and local leadership. "Think globally, but act locally" as environmental advocates have always said.

10.1   Denis Hayes: Eco-Elder

Figure 10.1 Denis Hayes. Public domain.

Denis Hayes, born in 1944, achieved prominence in 1970 when he co-ordinated the first Earth Day Event, which drew an estimated 20 million people around the world. Earth Day led to major environmental policy changes and *Time* Magazine named him "Hero of the Planet" in 1999. Today, in his eighth decade, he isn't finished and he remains a voice for environmental action on every front.

"It's a universal truth that people are most easily mobilized when there is a clear enemy," said Hayes. "In the late sixties, it was pretty clear who was causing air and water pollution, because we saw smokestacks pouring into the atmosphere."

Hayes remembers how passage of the Clean Air Act can give lessons for climate advocates today: "The Clean Air Act was an imperfect vehicle, but it did dramatically change air quality," he says. "You can have something that looks elegant on paper, but if you can't pass it, it doesn't do you any good. Take the imperfect thing and move forward. We've got to create momentum that says, 'We're chipping away at this.'"[10]

Hayes urges people today to maintain hope and to pay attention to the profound, lifesaving benefits of climate action right now.

Climate and Aging Groups

Third Act

Third Act is a national organization mobilizing people over 60 to take action about climate change. It is reimagining what elderhood might mean for countless numbers of Americans: to protect the climate and to strengthen democracy. It was launched in 2021 by environmentalist Bill McKibben whose goal was clear:

> We started organizing Third Act because we started to understand how much power those of us over the age of 60 possess . . . There's a lot of us, 70 million people over the age of 60. . . We punch way above our numbers politically, because we all vote. And, fair or not, we ended up with most of the money. Baby Boomers and the Silent Generation have about 70 percent of America's financial assets.[11]

Robin Kimmerer, author of *Braiding Sweetgrass*, said that in the Potawatomi Native American tradition, "elderhood is not a time for stepping back, but for stepping up, for stepping into your own power . . . Becoming an elder is both a precious gift and a serious responsibility—a responsibility for sharing knowledge, and for safeguarding the future of life."[12]

When Bill McKibben started Third Act he admitted that he didn't know if it would work. But, within just a year, it quickly grew to 50,000 members around the USA. Third Act soon organized itself into active Working Groups spread out across the country, representing geography (e.g., Third Act Sacramento, Third Act Ohio) and also vocation: for example, Third Act Educators, Third Act People of Faith.

Third Act's goal has been to ramp up the pressure on the fossil fuel industry. Third Act believes that much of the action needs to move away from Washington toward Wall Street, an approach reminiscent of Occupy Wall Street in earlier years. The name "Third Act" suggests a time in the 1960s when young people accelerated women's rights, equality, and political power around the civil rights movement. The first Earth Day in 1970 marked a high point for environmental advocacy. The idea is that the Silent Generation and Boomers today, who hold the bulk of the country's wealth, could pressure financial institutions to stop funding the fossil fuel industry. Third Act seems to be trying to engage the over-60 crowd in ways that Fridays for Future has targeted teens and school kids."[13]

A second goal of the Third Act is to ensure that democracy remains strong in the USA. This goal requires standing up against voter suppression laws at all levels of government. Showing up at the polls is only part of the picture. The Third Act also aims to push climate policies in public utilities and to develop stronger local resilience through mutual aid.

Above all, the Third Act wants to help redefine what it means to be an older American, in continuity with the generation of the sixties in which people see themselves as change agents over the whole life course. In his PBS presentation, Bill McKibben said, "The nearer we get to the exit, the more we understand that legacy is not a completely abstract idea. Legacy is the world you leave behind for the people you love most."

Some of Third Act's ambitious goals have already been achieved. Third Actors, as its members called themselves, mobilized in large numbers to help pass the Inflation Reduction Act (IRA) and Heat Pumps for Peace. By most accounts, they had a significant influence in the 2022 elections. This action in legislative politics was in keeping with Third Act's goal of preserving American democracy.

Third Act has also emphasized the importance of intergenerational collaboration.[14] In their view, elders need to step up to the plate: "So far the kids have had to do all of the work and they've done an amazing job, but it's not fair to ask 18-year-olds to solve this problem," said Bill McKibben, founder of Third Act: "Older people have got money and structural power coming out of our ears. We have to show young people we have their back."[15]

Intergenerational collaboration is key at the ballot box. Every year, 4 million Americans turn 18 and become newly eligible to vote. But less than half of them do vote because they are not registered. Third Act has allied with the Civics Center in a program called "Senior-to-Senior." It brings elders together with high school seniors to help make sure that younger people turn out to vote.[16]

Third Act has worked to strengthen national climate legislation. For example, Bill McKibben Act played a role in defeating West Virginia Senator Joe Manchin, who wanted a special "side deal" following up on the IRA, the landmark law against climate change. Senator Manchin's proposed side deal was eventually withdrawn, and his effort was defeated.

But McKibben also acknowledged a point made by Manchin in favor of permitting reform, a point climate advocates will need to hear. Permitting reform will prove necessary to build electrical transmission lines across America. The biggest factor for projected emission under the 2022 IRA will be electric power generated by expanded publicly owned utilities.[17] Individual consumers can do little on their own to create such utilities but expanded publicly owned utilities will be on the policy agenda for the future. Another example of a familiar imperative: think globally, act locally.

McKibben has repeatedly emphasized the importance of young and old working together. The Third Act elders played its part but so did the Sunrise Movement, which mobilized young people. In McKibben's words: "If the young people of the Sunrise Movement did the lion's share of the work to get the IRA passed, older people have played a not-insignificant role in blocking this side deal."[18]

Elders Climate Action

ECA is a non-partisan movement of elders committed to making the voices of older people heard on the climate crisis.[19] It is a division of Elders Action

Network, a group which includes efforts on behalf of democracy and social justice. In their own words, these elders are acting

> For Our Grandchildren, Future Generations and All Life. We are elders, including grandparents, great aunts and great uncles who care about the future for all children. As ECA members, we are determined to do all we can to leave a sustainable planet for future generations.

Elders Climate Action

The basic goal of ECA is to change national policy to avoid further damage to the earth's climate. The effort begins with the understanding that older Americans vote at a higher level than other age groups, so they are for that reason alone in a position to influence policy. ECA works from these core policies and principles through concrete efforts such as its Action Tool Kit, webinars and videos, links to other climate advocacy groups, and intergenerational programs that begin with grandchildren and encompass other younger people engaged to take action with ECA. The organization hosts multiple chapters around the country, sponsors events and action programs, and hosts national calls and other outreach steps to mobilize its members.

These efforts begin with the awareness that, in contemporary America, elders' talents and experiences are largely untapped. Wisdom gained over decades of living largely lies fallow and neglected. In this respect, ECA represents one form of "productive aging." Along with later life employment, this political action is one way of avoiding an unfortunate waste of invaluable human resources that can be applied to addressing today's societal and environmental problems.

Beyond specific climate policies, Elders Action Network aims to initiate a cultural shift so that elders reclaim a place in providing learning, wisdom, and guidance for the wider community. Instead of acting merely to promote health or income security, this effort would position elders to become social movement collaborating across generations for a better future for all.

Even beyond this humanistic goal, such a movement aims at a universal narrative including all living beings, interconnected within a larger unity. Climate justice then becomes part of a search for enduring meaning. Today's planetary crisis threatens ecological and cultural wellbeing. But ECA identifies the crisis as a consequence of humanity's lost reverence for life. The crisis, then, is also an opportunity to reimagine the future. Like other environmental groups, ECA seeks to remember the legacy of indigenous ancestors along with insights from wisdom traditions around the globe.

Younger generations sometimes feel inclined to blame elders for leaving them a planet in ecological crisis. ECA recognizes that elders are part of the problem but now aims to become part of the solution. The group is inspired by Ghandi's famous maxim: "Become the change you want to see." For this reason ECA encourages service activities by members and educates its members and other elders on how to change activities of daily living, including investment and consumer behavior so that every step helps elders to "walk the walk" of the principles and policies urged by ECA itself.[20]

ECA's community nearly doubled in 2022. In 2023, with COVID less threatening, ECA was able to convene a successful retreat event, built around the theme of *The Good Ancestor* book.[21] One element in their strategy is serious attention to holding the financial sector accountable for fossil fuel investments. Third Act gave major attention to the role of banks, but ECA has developed a strong partnership with Fix My Funds and other groups concerned about investment for a green retirement. It works to make sure its members know how to take advantage of the benefits under the IRA, the Infrastructure and Jobs Act, and the White House's Justice 40 Initiative.

ECA is part of a family of initiatives under the rubric of Elders Action Network, which includes parallel groups such as Elders Defending Democracy and Elders for Regenerative Living. ECA works collaboratively with Third Act and has been eager to forge intergenerational ties with youth groups like the Sunrise Movement. ECA has worked through a range of initiatives.

Membership

ECA's messages on social media and by email are now reaching over 26,000 followers.

Political Advocacy

ECA in 2022 sent more than 12,000 email messages to Members of Congress and the Biden administration. The group has developed a close alliance with Climate Action Now, a cell phone app that enables citizens to take action and send messages, with more than 100,000 contact during that year. ECA has worked closely with the Environmental Voters Project and with Climate Action Now in this approach.

This political mobilization helped assure the passage of the landmark Inflation Reduction Action ($369 billion) and the Bipartisan Infrastructure Law, both promising historic levels of investment in responding to climate change.

Financial Advocacy

ECA became a partner in the Stop the Money Pipeline Coalition, pressuring big finance companies to fund fossil fuels developments and infrastructure.

Collaboration

ECA was a founding member of the Breathe Again: Healthy Air is Healthcare Collaborative, a partnership with nine climate organizations providing tools, resources, and training about the impact on health from climate change and air pollution.

ECA also entered into partnership with the Environmental Voter Project, launching an "Elders Promote the Vote" campaign co-hosting weekly phone banks in which ECA volunteers contacted tens of thousands of voters ahead of the midterm elections.

Elders Action Network: Take the Pledge

In 2023 Elders Action Network mobilized its members to take a pledge for climate action on one—or even all five—of the following activities:

- Buy fewer new clothes. Say no to fast fashion and get creative with vintage and alternative choices.
- Eat a plant-rich diet with minimal food waste.
- Make fuel-efficient and less polluting transportation decisions. Make fewer transportation-intensive purchases by traveling less and eating locally grown, in-season food.
- Reduce electronic waste by not purchasing all the latest electronics and electrical appliances and repairing or replacing the ones you have only when they no longer do the job.
- Consider or even commit to other, longer-term significant life changes, for example, make my next car electric, participate in Elders Action Network initiatives such as Joining a Climate Action Chapter, Electrify Everything, and Eliminate Plastics.

Gray Is Green

Gray Is Green, the National Senior Conservation Corp, was founded in 2008 to guide members of America's older generations interested in environmental sustainability.[22] Gray Is Green serves as a clearinghouse for older adults interested in greening their lives, learning about sustainability, advocating for climate change policy, and serving as a resource for younger people concerned with sustainability.

In 2007 a group of retired professors from Yale University interested in environmental conservation decided to "green" their retirement home in Hamden, Connecticut. Professors Robert Lane, Arthur Galston, and other retired faculty began to organize conservation resources. They were quickly shocked to discover there was nothing aimed at enlisting older people. They realized that their own generation was largely responsible for environmental degradation, and they understood that seniors have significant political and financial resources, so the created Gray Is Green to change the situation.

The National Senior Conservation Corps publishes a monthly newsletter and provides resource materials to seniors interested in advocating for policies to address climate change and other environmental issues in their own communities. Older adults interested in learning how to prevent global warming and help the environment are provided with communications and curriculum and can also participate in an environmental speakers bureau. For instance, one community in California installed a photovoltaic generator to convert sunlight into energy. Another in New Jersey arranged for an energy audit to significantly reduce its own energy consumption.

Gray Is Green has developed a partnership with Recyclebank, an environmental organization which aims for waste-free communities through a unique rewards-for-recycling program. It also provides a platform called Storyboard, which features environmental themed stories by seniors. Seniors reflect on life experiences, connections with nature, and a shared sense of responsibility toward the environment. More recently, Gray Is Green published an e-book on senior environmental and climate action and advocacy, a publication now available, without charge, to multiple groups and individuals around the country.

In September 2014 Gray Is Green was part of the People's Climate March in New York City. The National Senior Conservation Corp joined with 1,500 other organizations, including labor unions, churches, schools, and community groups in this large-scale activist event convened by the People's Climate Movement to advocate global action against climate change. It is estimated that over 300,000 participants were there, making it the largest climate change march in history. The event was conceived as a response to the U.N. Climate Summit of world leaders, which took place in New York City just two days after the People's Climate March.

The philosophy of Gray Is Green is inspired by the belief that change in climate policy is possible, but it requires people from all walks of life to be part of the solution.[23] The e-book, *A Practical Guide for Environmental Eldership*, gives guidance on how to take steps take will make a difference in climate change.

10.2 Robert Lane: Eco-Elder

As a young student at Harvard (class of 1939) Robert E. Lane was already an activist, alarmed by Kristallnacht and the persecution of Jews in Nazi Germany. He soon became a national leader in refugee work, recognized by a personal letter to him from President Franklin Roosevelt. After wartime service, he earned a Ph.D. in political science and served as Chair of Yale's Political Science Department, becoming an international leader in that field.

As he grew older Robert Lane saw deeply into the political economy that had commanded his attention. He argued convincingly that, despite economic growth, there is a loss of happiness in market democracies.[24]

Once beyond a minimal poverty level, getting more money doesn't make us happier. It happens at the same time there is a tragic loss of family solidarity and community integration, including distrust of others and of the political order. The threat to democracy seen in the USA, as well as other rich countries proves his point. It is not a surprise that Prof. Lane, allied with others, would founded an environmental organization.

When Lane was 86 years old he became alarmed by climate change: "I was alarmed by the implications of global warming and I was moved to action." Retired from Yale University, Lane and his wife moved into a retirement community in nearby Hamden, Connecticut. In the retirement community Lane and others found no central organization bringing older together around climate change, so this group of Eco-Elders created their own.

Eventually, Lane led the way for residents in 50 other retirement communities across the country to form the Gray Is Green, National Senior Conservation Corps.

Robert Lane's work with Gray is Green was recognized with a Connecticut Governor's Climate Change Leadership Award in 2008 as well as an Encore.org Purpose Prize Fellowship in 2009. For full disclosure, I note that I have served for some years as chair of the Board for Gray Is Green, and I was able to meet and get to know Prof. Lane in his retirement community, as well as to see my own then-retired biology professor, Prof. Arthur Galston, also active in Gray Is Green. Both of them in retirement represented a commitment to lifelong learning and to activism on behalf of generations that they knew they would never live to see. Prof. Lane died in 2017 at the age of 100.

American Association of Retired Persons

AARP, originally the American Association of Retired Persons, is the largest organization of older people in the world.[25] It is not an aging climate advocacy group, but it plays an important role in all aging policy issues. With 38 million members AARP membership represents 17% of the voting public across all political, religious, and geographical divides. This figure underestimates the importance of AARP because older voters vote with a higher turnout than any other age group. AARP membership is, roughly, one-third Democrat, one-third Republican, and one-third Independent. All religious and ethnic groups are represented.

In 2022 a movement was launched among the members of AARP to bring AARP into climate advocacy. The idea was that since AARP is not one of the "usual suspects" when it comes to the environment. For that very reason AARP engagement could be a political game-changer to help promote public policy around climate change.

With the largest circulation magazine in the country and with the highest voter turnout of any age group, AARP is a powerful lobby in contact with all political actors. AARP's position can be crucial. For example, in 2010, AARP came down in support of the Affordable Care Act. In a hotly contested struggle, AARP's position made all the difference, as I saw myself firsthand in 2010 when I was serving as Vice President for Academic Affairs for AARP in Washington and working for the Affordable Care Act.

Despite political diversity, all 38 million AARP members share one thing in common: in the next 10 years, all—without exception—will be facing the consequences of climate change on planet Earth. AARP members may not understand all the science involved, but all recognize that fire and flood, drought, and heat waves are already part of the "new normal." Those AARP members who are old today are facing a world different from the world we grew up in.

For all these reasons, it could be hoped that AARP might be in a position to overcome political polarization and help deliver the message that changing climate effects all Americans. There is reason to think that this hope is coming to pass. In June 2021 there was an unprecedented cover story, "Climate Change and You," in the *AARP Bulletin* magazine.[26] In that cover story AARP unequivocally stated that climate change is real, it is happening now, and showed how it is disproportionately impacting older Americans. On the state level, an even more powerful recent AARP podcast about the heat wave in California made it clear that the burning of fossil fuels is to blame, so government action is needed to solve the problem.

Thousands of AARP members have signed a petition, urging AARP to get more involved in climate advocacy, and that campaign is having an impact.[27] The petition campaign is solidly in favor of intergenerational collaboration, as it states: "We urge AARP to act in solidarity with young people who are striking, protesting, and otherwise calling attention to this looming catastrophe. We owe it to them—our children, grandchildren and all future generations—to use our influence and resources to assure them a world in which they can live safely."[28]

For example, in spring, 2023 AARP California became the first AARP State office to announce plans to launch a pilot climate project: "Electrify Your Savings". AARP understood that the IRA opens up an opportunity for AARP members to both save money and help move toward clean energy.[29] The IRA helps every homeowner, small business owner, renter, and others to save money on energy costs in the process. AARP's core mission—to protect the health, financial security, and livable communities of seniors and their families—is well aligned to address the climate threat.

Mobilizing AARP as an agent for climate advocacy it not simple and is not easy. Older people, as a group, are more likely to vote for Republicans and more likely to be exposed to news sources, like Fox TV, that tend toward denial of the climate crisis. When AARP published its big cover story on climate in 2021, it did receive pushback from readers and that pushback was recognized by national AARP staff. Pushback in itself need not prevent action.

When AARP supported the Affordable Care Act in 2010, that decision prompted one of the largest levels of people resigning from AARP in its history, although this fact has never been publicly discussed. That incident shows that AARP can and will act on significant policy matters even if it creates discord. In the case of climate change, the impact of climate is decisive for health, income, and livable communities—all top priority areas for AARP.

Other Climate Advocacy Groups

Extinction Rebellion

The Extinction Rebellion is a nonviolent direct action movement challenging inaction around climate change and the mass extinction of species which, ultimately, threatens the human own species itself. Originating in Great Britain, Extinction Rebellion and its strategy is inspired the civil rights movement and the nuclear disarmament movement. Its leaders have thought deeply about these past precedents and about the meaning of nonviolent resistance.[30]

The biggest traditional environmental advocacy groups, like the Sierra Club or the Wilderness Society, have always worked within the prevailing political system. But other groups, like Earth First, have been more radical and have pursued direct actions that can be disruptive and controversial.

Kim Stanley Robinson's book *The Ministry for the Future* depicts a world in the not-so-distant future, still the 21st century, when climate change has provoked disasters requiring unprecedented human action to stop the global catastrophe. Robinson opens his own book with a horrifying account of a heat wave in India, a heat wave where the wet-bulb (humidity) level remains so high that millions of people die. This extreme death toll, in Robinson's story, prompts the Indian government, to launch a geoengineering intervention putting chemicals into the atmosphere to reduce solar radiation.

In Robinson's novel, we meet the Children of Kali, a group engaging in acts of ecoterrorism against governments and individuals when those groups perpetuate global warming. For example, the Children of Kali shoot down passenger airplanes, causing the death of innocent people to make it unsafe

Figure 10.2 Extinction Rebellion logo. Public domain.

to fly and to command attention to climate change. They take every step possible to reduce fossil fuel use by reducing container ship transport and mining operations.[31]

Is it true that desperate times require desperate actions? One thinks of abolitionist John Brown (1800–1859) who fomented anti-slavery rebellion prior to the American Civil War. When has the time come to act in a more radical fashion? Consider these words of David Brower, lifelong environmentalist, written at the age of 79:

> At my age, they can't do much to hurt me. I have a new freedom. They can't change my career, I've got it made. If I go to jail or am executed, it doesn't matter—though I'd rather stick around. I'd rather be out of jail than in, but not at the expense of this new-found freedom. Young people don't have this liberty. They've got years ahead of them, families, need for income. They can't alienate themselves too much from the system. I say to people my age, "You have this freedom. Please use it. You've had a role in whatever's happened to the earth, and it hasn't been that good. You now have a role in doing something about it. If you're going to die, make sure your boots are on. There are so many of us . . . More and more of us . . . You've got to sound off. The older you are, the freer you are, as long as you last."[32]

David Brower was the long-time Executive Director of the Sierra Club, America's most influential environmental organization. Later, in his old age, he became the founder of another important advocacy group, Friends of the Earth. Brower died in 2000 at the age of 88, still laboring on behalf of future generations. It was Brower who loved to quote the African proverb "We don't inherit the earth from our ancestors, we borrow it from our children."

Roger Hallam, the founder of Extinction Rebellion, is now in his mid-fifties, not quite an elder, but past middle age. He is the author of *Common Sense for the 21st Century: Only Nonviolent Rebellion Can Now Stop Climate Breakdown and Social Collapse*, whose title well expresses the mission of Extinction Rebellion. Hallam believes that traditional approaches—"working within the system"—are past the time of effectiveness. Radical action is called for, and he cites as examples Suffragettes and Martin Luther King as examples. More recent examples are Occupy Wall Street and Black Lives Matter. Nonviolent direct action is what is called for, Hallam believes.[33]

How much power do elders actually have to force changes in response to the climate crisis? Older people are a huge group. Americans over age 65 now number more than 50 million, more than 16% of the population. Extinction Rebellion cites the work of Harvard political scientist Erica Chenoweth about the proportion of people needed to make a change in society. Chenoweth calls this the 3.5% rule of social change. She studied hundreds of demonstrations across the 20th century and found that nonviolent protests are twice as likely to succeed as armed conflicts, and protests that engage at least 3.5% of the population have never failed to bring about change. Again, elders alone make up more than 16% of the American population. Will extreme weather events, along with convincing public protest, change attitudes and affect public policy?[34]

For Americans and other countries, too, the key is likely to come from alliances that reach across ethnic lines and across age group lines: in short, inclusive, intergenerational activism.

Like the Third Act in the USA, in the spring of 2023 UK Extinction Rebellion convened large public protests, but with uncertain outcomes.[35] There remain unresolved questions about how to conduct a public protest in support of action around climate change. In 2019, for example, thousands of members of the British branch of the Extinction Rebellion group launched a demonstration on the London subway system (The Tube) designed to bring transit to a stop and gain public visibility around the climate crisis. The effort caused large disruption but also resulted in public animosity because so many lives were disrupted by the protest:

> "At the end of the two-week global 'uprising', members of the movement's political circle announced that it needed to learn from the angry scenes at Canning Town tube station last Thursday when commuters dragged protesters from the roof of an underground train and set upon them. Eight XR activists were arrested during the disruption, joining a total of 1,768 held during the fortnight of demonstrations . . ." Senior figures in Extinction Rebellion have been forced to rethink future tactics. But despite the controversy over some tactics, many believe large-scale disruption remains effective in applying pressure on the government to tackle the climate emergency.[36]

Launching public protests and demonstrations is a challenge for leaders of a protest to adopt what Tibetan Buddhists call "skillful means" to achieve the goal and not to frustrate the purpose by creating public antagonism toward the protesters. The climate crisis may not happen at the right speed or pace in order to be seen as a crisis. Kim Stanley Robinson, author of *The Ministry for the Future*, cites the poet Gary Snyder, who spoke about this idea of the speed of a crisis.

The *Ministry for the Future* opens with a catastrophic heat wave in India, which sets the stage for what will follow. The book is fiction, "cli fi." But we already live in a world where heat waves are already happening. If an event is happening fast—whether the war in Ukraine or a violent storm—then we have no time to pay attention to intellectual debates: we're just trying to survive. But if the crisis is happening too slowly—over thousands of years—we may not respond because we adapt to a "new normal" and then stop noticing it. It may be that the climate crisis today is happening at a speed where we can recognize events—such as extreme weather events—if only we can "connect the dots" and realize that this "new normal" is actually a crisis unfolding before our eyes. Today's elders remember earlier decades in their lives can recognize the climate crisis and help to warn others, both those of their own age and younger generations. All generations need to become agents in respond to climate change.

Climate Action Now is an app downloaded to cell phones that enables individuals to take climate actions every day, such as sending messages to public officials (The U.S. President, Congressional representatives, and state legislators) as well as media managers and corporate leaders (CBS, USA Today Home Depot, Citicorp, etc.). More than a million messages have been sent since 2022.[37]

Citizens Climate Lobby is a nonprofit group that empowers people to work on climate policy. Supporters are organized in more than 420 chapters across the USA building support in Congress for a national bipartisan solution to climate change. Their signature goal is putting a price on carbon pollution to cut America's carbon pollution in half by 2030.[38]

The **Environmental Voter Project**'s goal is to identify inactive environmentalists and transform them into consistent voters to build the power of the environmental movement. They estimate that over 8 million environmentalists did not vote in the 2020 presidential election and over 13 million failed to vote in the 2022 midterms. EVP will call, canvass, mail, or send digital ads to millions of low-propensity environmental voters each year with just one goal: turning them into better voters.[39]

The **Climate Reality Project** is a nonprofit organization devoted to education and advocacy on climate change. It was founded in 2006 by former Vice President Al Gore as the Alliance for Climate Protection to mobilize civic action against climate change. Gore provided funding from money earned through his acclaimed documentary film *An Inconvenient Truth*.[40]

The **Post Carbon Institute** is a think-tank producing research and publications around climate change, energy policy, and related questions about sustainability and long-term community resilience. Founded in 2003, the Post Carbon Institute has become an important source for policy analysis around climate change.[41]

The **Sunrise Movement is** a youth-based apolitical action group advocating on climate change since it began in 2017. Allied with groups like Fridays for Future, Sunrise has mobilized young people and others of all generations in support of the Green New Deal and other efforts on behalf of climate sustainability.[42]

350.org is an international environmental group devoted to the climate crisis. Its goal is to stop fossil fuels and promote a transition to renewable energy through a global, grassroots movement. Its name signifies that 350 parts per million of carbon dioxide is the upper limit for global carbon pollution. The group promotes online campaigns, grassroots activities, and mass public actions, such as its leadership in the Global Climate Strike 2019.[43]

Fridays For Future is an international, intersectional movement of students striking for the climate. It began in August 2018, when 15-year-old Greta

Thunberg sat in front of the Swedish parliament every school day for three weeks to protest the lack of action on the climate crisis. Since then, millions of people have participated in Fridays For Future strikes around the world.[44]

The New Politics of Age and Climate

Let's start with some clear facts. For the remainder of the 2020s, every day 10,000 people will turn age 65 and they'll continue to grow older every single day after that. In short, population aging continues to advance even if it doesn't grab the headlines. But extreme weather events do grab headlines. By all estimates, threats from climate change—the Four Horsemen of the climate apocalypse—fire, flood, heat wave, and drought—will also continue during this period. Elders and climate both proceeding at the same time. The question: Can older people be part of the solution?

Denial is not just a problem for climate deniers. It's also a problem for climate advocates. For example, I came across an appeal from a climate advocacy group urging more older voters to get to the polls in support of climate work: "We need to pull more elders into the fight to push for the changes that are so desperately needed."[45] This appeal sounds good until you look at the facts: how do older voters actually vote compared to younger voters? The answer may not be what we expect.

Beginning in 2016, when Donald Trump declared for President, I started warning gerontologists and aging advocates about the danger he posed. I pointed out that older voters were supporting Trump in larger numbers than other groups. Sure enough, in 2016 and 2020, according to reliable exit polls,[46] older voters voted for Trump at a higher rate than younger people. In 2016 and 2020, voters eligible for AARP (i.e., age 50-plus) voted for Trump at a rate higher than any other age group (majority of 52%). By contrast, younger voters were the decisive age group supporting Biden in 2020, although their turnout was lower than for older voters, as has often been the case. Perhaps those of us elders who care about climate change need to mobilize younger people and make sure that turnout is effective.[47]

When I pointed this out to colleagues in the field of aging, you could immediately notice denial: their eyes glazed over, listeners looked the other way, and they quickly changed the subject. It was all just too much to bear. If we spend our lives depicting elders as victims, it's hard to think of them as villains. No wonder denial is a common response.

But there is always hope for change, and even some evidence for it. Voters aged 65 and older are second only to those between the ages of 18 and 34 in naming climate and the environment their highest political priorities.[48] The unanswered question is whether such sentiments and survey results will translate into voting behavior when climate change is at stake. A report from the Environmental Voter Project says "gray is the new green," suggesting that voters who prioritize climate are numerous enough to swing elections in key states.[49] This is clearly a sign of hope.

Maybe victims-or-villains is the wrong way to think about this problem. An important book on environmental policy is titled *Strangers in Their Own Land*.⁵⁰ That title gives a hint about why senior voters have supported Trump. For elder climate advocates, like myself, this is not good news. Denial is tempting in thinking about this challenge. But denial is a bad way to go, whatever our political opinions. In the past (2016 and 2020) senior voters were more likely than other age groups to vote for Trump. But, as a mutual fund prospectus always says, past performance is no guarantee of future results. More recently, voters aged 65 and older have become second only to those between the ages of 18 and 34 in naming climate and the environment their highest political priorities.

Local Elder Activists

Sharon Lavigne, now age 70, spent 38 years teaching in public schools in the Mississippi River town of St. James, Louisiana. It's the region depicted by Arlie Hochschild in *Strangers in Their Own Land*, an area named "Cancer Alley" because of the prevalence of disease prompted by petrochemical plants and their dangerous pollutants. But now some limited-income minority residents are fighting back, and Sharon Lavigne is one of them. In 2021, she won the Goldman Prize, a coveted award for environmental activism. The Prize recognized her work in stopping a proposed plastics factory that would have cost more than a billion dollars.

John Beard, in his mid-60s, lives in the east end of the Texas Gulf Coast, in the middle of petrochemical plants where Beard himself worked for 40 years. He is now the founder of the Port Arthur Community Action Network, and he believes that experience matters on the front lines of environmental justice. "They say every generation has its challenges, but I think we probably got the greatest challenge," said Beard. "Our challenge is to save humanity, to save this planet, which is the only planet we have. If we destroy it, we are in essence destroying ourselves."

We see the same pattern when municipal waste incinerators produce toxic pollution and leave waste clouds thick with deadly health implications, as Earth Justice has noted (April 2023), adding that most are in communities of color.

Sharon Lavigne and John Beard are not card-carrying members of Third Act, the Sierra Club, or any other national organization. Instead, they're both like the hero in the film *Network* (1976). They just found themselves saying, "I'm mad as a hell and I'm not going to take this anymore." Lavigne and Beard were not strangers in their own land but activists on the front lines of environmental justice where age is no limit.

In the film *The Graduate*, Dustin Hoffman's character is advised that the key to his future is found in one word: "Plastics." It wasn't true. Elders of the generation of John Beard and Sharon Lavigne grew up listening to the DuPont company's tagline, "Better living through chemistry." That wasn't true either. But they both discovered that plastics and chemistry were what would destroy their local communities, and they acted when age has no limits. Environmental justice depends on it.⁵¹

10.3   Environmental Heroism Has No Age Limit: Eco-Elders[52]

Both Sharon Levigne and John Beard, Jr, are examples of African-Americans as Eco-Elders working on behalf of their own community and on behalf of us all.[53]

There are plenty of older Americans who understand that climate change is real and who also understand that Donald Trump was wrong when he called it a "hoax." African-Americans may be less likely to be deceived when people tell them that their suffering is an illusion.[54] Older people also understand that younger voters can be likely allies in a time of climate crisis. Intergenerational collaboration is the way to go, as this book has argued repeatedly. Older people need to reach out and mobilize those who will inherit the world we are leaving behind them.

False hope gets in the way of genuine hope. It's been said about Boomers in old age, "We are the people we've been waiting for," in the lyrics of a song by Bono. The words are supposed to celebrate a vision of progressive political victories dating back to the sixties. The reassuring message is this: don't be discouraged by defeats and setbacks. Demographic change—aging Boomers—will come to the rescue again.

But this optimistic slogan is not an accurate picture of the 1960s at all. The leaders of progressive cultural changes we summarize as "The Sixties" were not Baby Boomers but figures like Gloria Steinem (born 1934) or Martin Luther King, Jr (born 1929)—in other words, much older than the oldest Boomers. What about aging Boomers today? Well, there are plenty of aging Boomers engaged with climate advocacy groups such as Third Act. But, as an age group, Boomers voted for Donald Trump (a Boomer, born in 1946) at a higher level than any other age group, in both 2016 and 2020. As a group, Boomers were not all at Woodstock or at Earth Day. Instead, Boomers were more likely to be in the military and have served in Vietnam. Today aging Boomers are the backbone of Republican voters. Older voters, as a group, have not proved to be a friend of environmental issues, including climate change, in recent elections. That's simply a fact that must be recognized if climate advocacy can be effective in years to come.

John Della Volpe, director of polling at Harvard's Kennedy Institute of Politics, noted that, if the 2022 elections had been decided by voters 45 and older, then the Republicans would have won the House by a big margin and perhaps won the Senate, too. In that election, as in others, younger voters were the key. As it turned out, the 18-to-29 age group in 2022 had the second-highest turnout in midterm elections in nearly three decades.[55] There's an irony here in that younger voters, not older voters, made the difference. In short, it was the young, not the old, who protected democracy as well as the biggest national climate initiative signed by President Biden, the oldest man ever to occupy the presidency.

The great psychologist Erik Erikson wrote about old age as a time when people face ego-integrity or despair. In his own old age, he published a book titled *Life History and the historical Moment* (Norton, 1975). The historical moment for elders today looks back to a time when the USA saw success in preserving democracy in WWII and in the Cold War. But in a time of climate change, what will the future bring?

> In the 20th century, the United States invested in democracy around the globe to confront the threat of totalitarianism. Now, in the 21st century, the gravest threat to democracy around the globe is, I believe, the weakening of democracy here at home. It is time to bring the same commitment to securing our civic strength in the United States as we brought to fighting totalitarianism in the 20th century.[56]

What's this got to do with climate change? Everything. Third Act and ECA, to name only two groups, are making it a priority to preserve democracy. That means preserving openness about the future, which is another word for hope. Environmentalist David Orr put it best: "Hope is a verb with its sleeves rolled up." Act wherever you can: "Feed the opportunities, starve the problems." (Peter Drucker's words). The best advice is: don't go with a news fast, don't give in to denial of darkness but instead turn toward the light, even if only with incremental steps.

Young people can be a big part of climate advocacy. The American Climate Corp is an example of mobilizing youth to help fill a national shortage of skilled workers required for decarbonization:

> Beginning in the summer of 2024 the American Climate Corps is sending 20,000 18- to 26-year-olds across the country to install solar projects, mitigate wildfire risk, and make homes more energy-efficient. President Biden's New Deal-inspired program is modeled after Franklin D. Roosevelt's Climate Conservation Corps and attracted 100,000 applicants. As it rolls out, the climate corps continue(s) to draw criticism from the left for low wages and ageism, and from the right for being a "made-up government work program . . . for young liberal activists."[57]

Elders working with youth can be the key to generational justice for all:

> Human communities flourish on the diverse qualities that each age group brings and reflects for the others. Elders provide ancestral wisdom, conflict resolution, unconditional love, and a profound sense of presence that complement the attributes of adults, youth, and children. Without elders, there is a gap in the transmission of knowledge of how to thrive with the Earth, and a weakening of the intergenerational bonds that hold communities together.[58]

The IRA

The bill approving the IRA was a surprise. It wasn't the "Green New Deal," and it passed by the slimmest possible votes in both the House and Senate. But no one should underestimate the brilliance of the politics. The law is framed in a very different way than the language of sacrifice or even regulation. The IRA was an example of compromise, and that is not a bad thing. We need compromise because we need results.[59] The IRA of 2022 carries "reducing inflation" in its title. From a political perspective, who can be against cutting higher inflationary prices? In the IRA any tax increases are reserved for corporations, not for ordinary American taxpayers.

The IRA is also legislation about aging, appropriate for climate change in an aging society. A major part is devoted to reducing prescription drug prices for seniors on Medicare. The IRA, whatever its limits, must count as a politically brilliant way of linking age to climate advocacy. Whatever the views of senior voters about climate change in the past, they will now appreciate cutting drug costs.

Robert Reich makes the important point that the IRA represents a major policy shift away from the regulatory state of the 1930s to the 1970s, toward the subsidy state beginning in the 1980s. Reich asks, why this shift? He answers, "because of the change in the balance of power between large corporations and government. Today it's politically difficult, if not impossible, for government to demand that corporations (and their shareholders) bear the costs of public goods."[60] The discussion of climate justice—for example, under the impact of the IRA—sometimes takes account of race and ethnicity—but rarely does it take account of aging and the life course.[61]

Throughout his Presidency and into the 2024 election, Joe Biden has been criticized for his age. It is therefore an irony that one of his greatest accomplishments was the passage of the IRA—the largest investment in climate change in American history. Perhaps it would be going too far to say, with Justice Brandeis, that states are "laboratories for democracy." But sometimes they can be. That is the strength of a federal system of government. When action is limited at one level, it remains possible at other levels. When Congress is polarized or paralyzed, action on climate can shift to state and local initiatives. Funding from the IRA is becoming available to the states, and now governors and in many states legislatures can make climate action part of their initiatives for the future. The IRA provides more than $50 billion to states, and public policy at the state and local levels will play a bigger role even when action is paralyzed at the national level. The transition to clean energy cannot wait for national policy to take leadership. We live in a time when bottom-up action is also essential.

IRA has also been a major part of becoming a climate-conscious consumer:

> For all that has been written about the IRA, the most salient fact about
> it remains widely underappreciated. What is significant about the bill

is not just that it sends an enormous amount of money toward climate solutions, but that the money is almost entirely uncapped.[62]

The point is that the amount of government money in the IRA isn't specifically limited but depends on the demand for tax credits. The more people who ask for them, the more gets spent. What are the numbers here? The Congressional Budget Office estimates that the IRA will spend $391 billion, but a Goldman Sachs report put the number as close to $1.2 trillion. New York Hotel tycoon Leona Helmsley once said that only a few people pay taxes. If the "little people" apply for tax credits under the IRA, it could have a big impact. But getting people to understand how to become climate-conscious consumers, and how to get a payoff, will depend on educating and mobilizing people.

Political Change Around Climate Is Not Hopeless

The success of the IRA, which came as a surprise, proved that hope was possible. But there are genuine obstacles to the politics of climate in an aging society. In America, as in other countries around the world, the rise of authoritarian populism has been part of serious political polarization. Polarization makes people think that it's impossible to change political opinions, around climate change as around other subjects. However, some experiments in social psychology suggest that it may be easier than we think to change someone's political views, which is a good sign for climate advocates. Climate advocates need to adopt what Tibetan Buddhism calls "skillful means" in communicating with voters about climate change. As the researchers in a recent study concluded:

> There is no quick fix to the current polarization and inter-party conflict tearing apart this country and many others. But understanding and embracing the fluid nature of our beliefs, might reduce the temptation to grandstand about our political opinions. Instead humility might again find a place in our political lives.[63]

Survey research has shown that almost all Americans underestimate public support for climate policy. It's a good news–bad news story. It turns out that people estimate that 40% of Americans support climate policy changes. But the real support is between two-thirds and 80%: a big difference. In short, at least two-thirds of Americans believe that the government is not doing as much as it should to fight climate change. Public support is there, but people don't know it, and that ignorance may prevent them from talking about the subject and encouraging others on specific policy issues. This false social reality, in other words, has political consequences.[64]

The politics of climate advocacy has a parallel with the politics surrounding the political candidacy of former President Donald Trump. As in the past, Trump has a core of "base" supporters who don't care what crimes he committed or what he's done wrong. The base will vote for Trump no matter what. Many surveys suggest that this is a significant number of Republicans, as we saw in the 2020 election, which was closer than expected. But Trump has lost many votes among independent voters, as well as Democrats, of course. We do not need to convince Trump voters to defeat him at the ballot box.

On climate change, there is a small and dwindling group of voters who could be described as "climate deniers." Many are Trump supporters, of course. But the point is that it does no good to show such climate deniers that climate change threatens them or that alternatives like renewable energy are possible. It's not a good idea to expend energy trying to reach voters who can't be persuaded. Arguments are irrelevant. They are simply climate deniers, and it's a waste of time to try to convince them otherwise. But significant public opinion in America has shifted to recognize that climate change is anthropogenic: that is, it's caused by us and that there are things we can do about it. We're winning that argument already.

In thinking about citizen action for climate change, it's useful to look at the successes and failures of recent political efforts in the USA.[65] One of the biggest successes was Earth Day in 1970, which resulted in major policy changes in the U.S. government. Another success was the Montreal Protocol of 1987, an international treaty to protect the ozone layer. It's not irrelevant that in the summer of 2022, when the big climate law was finally passed, the USA was experiencing a time of tremendous wildfires, drought, and other extreme weather events. It's hard to get voters to focus attention on climate prospects in the future: They experience weather on a day-to-day basis, and it is the weather, not climate, that commands their attention.[66]

The real danger is not climate deniers but climate doomers: that is, people who recognize the threat but feel there's nothing they can do about it: nothing collectively, and nothing as individuals.[67] The whole point of this book has been to show how the Four Horsemen of the Climate Apocalypse are coming for people of all ages, young and old. The whole point is that there are things we can do, collectively and individually, in response to this threat. Becoming a climate-conscious consumer and investing in a green retirement are things we can do as individuals. However, the marketplace will not be effective without government policy to reinforce individual action. Individual action and collective action are reciprocal, and both are necessary. This argument underscores the importance of efforts to register voters, build grassroots support for climate-conscious candidates up and down the ballot, and then ensure turnout at election time. In a federal system like the USA, it underscores working at all levels: local, state, and national government. "Think globally, act locally" remains the imperative today as always.

The new politics of age and climate will remain unpredictable in its details. But what we know is that today new groups like Third Act, ECA, and Gray Is Green have emerged and continue to do their work. The message of HERE NOW YOU HOPE will continue to be the imperative for citizen climate action in the future.

Notes

1 Sophia Naughton, "Two-Thirds of Americans Care About Environment as Much as Their Finances" *OnePoll Survey* (Sept. 2022) at: https://studyfinds.org/being-sustainable-saves-money/

2 The concept of learned helplessness was developed by Martin Seligman and co-authors, *Learned Helplessness: A Theory for the Age of Personal Control* (Oxford University Press, 1995). On climate change, see Nadia Said and Vivien Wölfl, "Impact of Constructive Narratives About Climate Change on Learned Helplessness and Motivation to Engage in Climate Action" (July 22, 2023) at: https://doi.org/10.31219/osf.io/fq2bt

3 From Edmund Burke: "To be attached to the subdivision, to love the little platoon we belong to in society, is the first principle (the germ as it were) of public affections." *Reflections on the Revolution in France.* The phrase "little platoons" is used in Robert Nisbet's classic *The Quest for Community* (1953) and also as in Yuval Levin's *The Fractured Republic.* See also Russell Kirk, "The Little Platoon We Belong to in Society" *Imprimis* (Nov. 1977);6:11 at: https://imprimis.hillsdale.edu/the-little-platoon-we-belong-to-in-society-november-1977/ Political conservatives, often more than progressives, have understood the importance of mediating structures. But some on the left, such as Saul Alinsky, well understood the importance of little platoons.

4 See Benjamin Bowman and Sadiya Akram, *Handbook of Young People and Environmental Activism* (Routledge, in press).

5 "The Montana Climate Kids' Lawsuit Has Energized Activists, Including This One" *Washington Post* (Aug. 27, 2023).

6 "Why Intergenerational Thinking Is Essential to Heal the Planet" at: www.yesmagazine.org/environment/2023/04/04/climate-change-intergenerational-thinking

7 Quoted by Lawrence MacDonald, *Am I Too Old to Save the Planet: A Boomer's Guide to Climate Action* (Changemaker Books, 2023).

8 "Understanding the Growing Radical Flank of the Climate Movement as the World Burns" at: www.brookings.edu/articles/understanding-the-growing-radical-flank-of-the-climate-movement-as-the-world-burns/

9 Tweet from Jeff Stein, from @JStein_WaPo (May 18, 2023).

10 From Sari Harrar, "The First Earth Day Changed the World. Here's What We Can Learn from It" *AARP* (Apr. 20, 2020) at: www.outsideonline.com/outdoor-adventure/environment/pollinator-pathway-native-gardening/

11 "Bill McKibben, Environmentalist, Founder of Third Act" Brief but Spectacular, *PBS News Hour* at: www.pbs.org/newshour/brief/430904/bill-mckibben

12 Cotton, Emma, "Bill McKibben Launches 'Third Act' to Rally Older Americans Around Climate Change" *VTDigger* (Sept. 2, 2021) at: https://vtdigger.org/2021/09/02/bill-mckibben-launches-third-act-to-rally-older-americans-around-climate-change/

13 Sally Pollak, "Seven Days" (Mar. 29, 2023) at: www.sevendaysvt.com/vermont/the-over-60-crowd-steps-up-with-bill-mckibbens-climate-action-group-third-act/Content?oid=37895972

14 "Elders Seek to Supercharge Climate Action" at: www.theinvadingsea.com/2023/05/02/climate-activism-baby-boomers-third-act-protest-banks/

15 "OK, Boomer, Your Turn. Older Americans Blockade Banks to Protest Fossil Fuel Financing" at: https://insideclimatenews.org/news/21032023/ok-boomer-your-turn-older-americans-blockade-banks-to-protest-fossil-fuel-financing/
See also: Bill McKibben, "Climate Change Is the Legacy of People Over the Age of 60. That's Why We Must Protest" *The Guardian* at: www.theguardian.com/commentisfree/2023/mar/27/older-people-climate-protest-banks-ipcc

16 https://thirdact.org/act/senior-to-senior-high-school-voter-registration-2023/

17 Dan Neumann, "Inflation Reduction Act Signals Consumer-Owned Power Will Be 'Driving Force' in Climate Fight" at: https://mainebeacon.com/inflation-reduction-act-signals-consumer-owned-power-will-be-driving-force-in-climate-fight/
18 "The Crucial Years" (Sept. 27, 2022).
19 www.eldersclimateaction.org/
20 Julia Coombs Fine, "Closing the Concern-Action Gap Through Relational Climate Conversations: Insights from US Climate Activists" *Climate Action* (Dec. 5, 2022) at: www.nature.com/articles/s44168-022-00027-0
21 Roman Krznaric, *The Good Ancestor: A Radical Prescription for Long-term Thinking* (The Experiment, 2021).
22 Full disclosure. For some years, I have been Board Chair of Gray Is Green.
23 "In This Together Climate Action" (Dec. 4, 2022) at: www.nature.com/articles/s44168-022-00027-0
   See also Marianne Krsasny, *Connecting with Your Community to Combat the Climate Crisis* (Comstock, 2023).
24 Robert E. Lane, *The Loss of Happiness in Market Democracies* (Yale University Press, 2001).
25 Frederick Lynch, *One Nation Under AARP* (University of California Press, 2011).
   Full Disclosure: I served as Vice President for Academic Affairs for AARP (2004 to 2013).
26 David Hochman, et al., "What You Need to Know About Climate Change: How It's Already Affecting Your Health, Home and Safety—and What You Can Do about It" *AARP* (June 1, 2021) at: www.aarp.org/politics-society/history/info-2021/climate-change.html
27 For the AARP petition campaign, see: https://supportaarpclimateaction.org/
28 https://supportaarpclimateaction.org/petition/
29 Suzanne Potter, "Group Helps Older Americans Adapt to Climate Change" *Public News Service* (Mar. 7, 2023) at: www.publicnewsservice.org/2023-03-07/senior/group-helps-older-americans-adapt-to-climate-change/a83342-1
30 Rupert Read, "Extinction Rebellion: I'm an Academic Embracing Direct Action to Stop Climate Change" *The Conversation* (Nov. 16, 2023) at: https://theconversation.com/extinction-rebellion-im-an-academic-embracing-direct-action-to-stop-climate-change-107037
31 Kim Stanley Robinsons, *The Ministry for the Future* (Orbis, 2021).
32 Studs Terkel, *Coming of Age: The Story of Our Century by Those Who've Lived It* (St. Martin's Press, 1996).
33 Chapter 14: "The Civil Resistance Model" in *This Is Not a Drill: An Extinction Rebellion Handbook* (Penguin Books, 2019). See also: Roger Hallam, "How to Stop the Climate Crisis in Six Months" (Sept. 4, 2021) at: www.youtube.com/watch?v=f5B9AZ_YKlw
34 Eve Andrews, "Interest in Civil Disobedience Has Reached a Mini Climate Tipping Point" *Grist* at: https://grist.org/protest/interest-in-civil-disobedience-has-reached-a-mini-climate-tipping-point/
35 Marc Hudson, "Extinction Rebellion Gave It 'the Big One' with a Four-Day Peaceful Protest—Now What?" *The Conversation* (Apr. 27, 2023) at: https://theconversation.com/extinction-rebellion-gave-it-the-big-one-with-a-four-day-peaceful-protest-now-what-204581
36 "Tube Protest Was a Mistake, Admit Leading Extinction Rebellion Members" *The Guardian* (Oct. 20, 2019) at: www.theguardian.com/environment/2019/oct/20/extinction-rebellion-tube-protest-was-a-mistake
37 www.climateactionnow.com/
   Is it too late to affect the politics of climate change? Not at all. Pessimism and complacency are the enemy. Each day I send 50 messages to political leaders, business executives, and media managers—all free, all composed to send out from

Climate Action Now volunteers. They recommend only five actions or messages a day: less than 5 minutes. But anyone can easily do more.

38 https://citizensclimatelobby.org/about-ccl/
39 www.environmentalvoter.org/
40 www.climaterealityproject.org/
41 www.postcarbon.org/
42 www.sunrisemovement.org/
    See also: "Inside 'The Very Secret History' of the Sunrise Movement" at: www.buzz-feednews.com/article/zahrahirji/sunrise-movement-climate-change-black-activists
43 https://350.org/
    See also: "It was the upstart that changed the face of America's environmental movement. But 350.org, founded by the legendary Bill McKibben, has been laid low by a budget crunch, equity fights and union strife." Zack Colman, "The Group That Brought Down Keystone XL Faces Agonies of Its Own" *Politico* (Feb. 20, 2022) at: www.politico.com/news/2022/02/20/350org-mckibben-boeve-keystone-00009866
44 https://fridaysforfutureusa.org/#:~:text=Fridays%20For%20Future%20is%20an,action%20on%20the%20climate%20crisis.
45 "We Won These Fights Once, and We Can Win 'Em Again" at: https://groups.google.com/g/goodneighbors/c/kfAXDHubjsM?pli=1
46 Pew Research Center: 2016 at: www.pewresearch.org/politics/2018/08/09/an-examination-of-the-2016-electorate-based-on-validated-voters/
    2020 at: www.pewresearch.org/politics/2021/06/30/behind-bidens-2020-victory/
47 Pew Research Center: 2020 at: www.pewresearch.org/politics/2021/06/30/behind-bidens-2020-victory/
    See also: "Why Older Citizens Are More Likely to Vote" at: https://money.usnews.com/money/retirement/aging/articles/why-older-citizens-are-more-likely-to-vote
48 "Gray Is the New Green: The Growing Strength of Older Climate Voters" at: www.environmentalvoter.org/sites/default/files/documents/2023-gray-is-the-new-green.pdf
49 *Gray Is the New Green: The Growing Strength of Older Climate Voters* Key findings from predictive models identifying environment-first voters aged 65 and older across 18 states. www.environmentalvoter.org/sites/default/files/documents/2023-gray-is-the-new-green.pdf
    See also: "Older Voters Are Second Only to Young People in Share of Climate Voters, New Study Shows" at: https://insideclimatenews.org/news/05122023/gray-is-the-new-green/
50 Arlie Russell Hochschild, *Strangers in Their Own Land* (New Press, 2018).
51 See "Pollution Harms People of Color the Most. How Biden's New Office Plans to Change That" at: www.usatoday.com/story/news/nation/2023/04/21/white-house-environmental-justice-office/11711790002/
52 **From: Craig Miller, "Environmental Heroism Has No Age Limit"** *NextAvenue* **(Apr. 20, 2023) at: www.nextavenue.org/environmental-heroism-has-no-age-limit/**
53 www.goldmanprize.org/recipient/sharon-lavigne/   https://biologicaldiversity.org/w/news/press-releases/john-beard-jr-honored-with-rose-braz-award-for-bold-activism-2021-12-08/
54 "Climate Action: Visionary Black Environmentalists Making A Difference" *EarthDay.org* (Feb. 26, 2024) at: www.earthday.org/visionary-black-environmentalists-making-a-difference/
55 John Della Volpe, "Republicans, Fear the Young" *NY Times* (Nov. 19, 2022) at: www.nytimes.com/2022/11/19/opinion/youth-vote-midterm-election.html?smid=nytcore-ios-share&referringSource=articleShare
56 Danielle Allen, "Pulled into Darkness, Turning Toward the Light Is Our Responsibility" *Washington Post* (Oct. 15, 2023).

57 "A New Climate Corps Will Turn Young People's Anxiety into Action" From: "24 Predictions for 2024" *Grist* (Jan. 3, 2024) at: https://grist.org/culture/24-predictions-for-2024-climate-trends/

58 From: "Bridging the Eldervoid: Can We Create a Society That Cherishes Its Elders?" at: www.yesmagazine.org/issue/elders-2/2023/11/30/community-bridge-wisdom
    ` See also: "Elder Power: First-Time Activists Show That the Front Line Isn't Just for the Young" at: www.yesmagazine.org/issue/elders-2/2023/11/30/first-time-activist-elder

59 For more on compromise, see "Solar Companies and Environmentalists Say They're Ready to Stop Fighting. They'd Better Be" at: www.latimes.com/environment/newsletter/2023-10-11/column-solar-companies-and-environmentalists-say-theyre-ready-to-stop-fighting-theyd-better-be-boiling-point

60 R. Reich, "How Biden Did It: From the Regulatory State to the Subsidy State" (Aug. 18, 2022) at: https://robertreich.substack.com/p/how-biden-did-it

61 Manann Donoghoe, Andre M. Perry and Hannah Stephens "The US Can't Achieve Environmental Justice Through One-Size-Fits-All Climate Policy" *Brookings* (June 1, 2023) at: www.brookings.edu/blog/the-avenue/2023/06/01/the-us-cant-achieve-environmental-justice-through-one-size-fits-all-climate-policy/

62 David Roberts, "Building a Movement That Can Take Full Advantage of the IRA" *Volts* (Apr. 26, 2023) at: www.volts.wtf/p/building-a-movement-that-can-take

63 Philip Pärnamets and Jay Van Bavel, "How Political Opinions Change" *Scientific American* (Nov. 20, 2018) at: www.scientificamerican.com/article/how-political-opinions-change/

64 Gregg Sparkman, Nathan Geiger and Elke U. Weber, "Americans Experience a False Social Reality by Underestimating Popular Climate Policy Support by Nearly Half" *Nature Communications* Aug. 23, 2022;13:4779.
    See also: "The Underestimated Climate Supermajority" at: https://insideclimatenews.org/news/17092022/warming-trends-a-comedy-with-solar-themes-a-greener-cryptocurrency-and-the-underestimated-climate-supermajority/

65 Alice Bell has documented that history in *Our Biggest Experiment: An Epic History of the Climate Crisis* (Berkeley, CA: Counterpoint, 2021).

66 Elke U. Weber, et al., "Constructed Beliefs in Climate Change Perception" (June 23, 2013) at: http://cred.columbia.edu/research/all-projects/heuristics-and-constructed-beliefs-in-climate-change-perception-effect-of-outdoor-temeprature-question-construction-and-cognitive-primes

67 Brian Kateman, "Climate Doomerism Is Dangerous. Climate Optimism Is Even Worse" Arguing That Hope Is Crucial in Effecting Change, While Blind Optimism Is Not, *Fast Company* (Aug. 23, 2023) at: www.fastcompany.com/90943444/climate-doomerism-is-dangerous-climate-optimism-is-even-worse

Part V
Hope

11 Fear and Hope in Climate Crisis

In a dark time the eye begins to see.

—Theodore Roethke

You do not need to know precisely what is happening or exactly where it is all going. What you need is to recognize the possibilities and the challenges offered by the present moment and to embrace them with courage, faith and hope.

—Thomas Merton

Waking From the Dream

James Joyce, through the voice of Stephen Daedelus in *Ulysses*, once said: "History is a nightmare from which I am trying to awake." Joyce died in 1941, as the WWII was just underway. But dreams of the horror were already apparent to many people.[1] We like to believe that our hopes and fears are what we recognize consciously. The topics discussed in this book fall into that domain, including both threats and actions to respond to the threats of climate change. But we are wiser to understand that the climate crisis also has features of a nightmare: "It can't be happening." This response is related to denial, and it is a coping mechanism embedded in our entire civilization, including both young and old.

The poet Theodore Roethke, influenced by Meister Eckhart and Martin Buber, understood that the darkness and despair of a situation can be the beginning of our ability to see: that is, to see things as they are: "In a dark time the eye begins to see."

An aging society means a society with more and more people who are old, by whatever measure we use. Population aging is the world in which we are living, and it shapes the history of our time. We are confronted by the question raised by Joyce: if this is our history, if this is our nightmare, how can we wake up? This is the context in which to consider fear and hope in a time of climate crisis.

DOI: 10.4324/9781003345992-16

Figure 11.1 "The Horsemen of the Apocalypse" a woodcut by Albrecht Dürer depicting Death, Famine, War, and Pestilence. Metropolitan Museum of Art, N.Y. City www.metmuseum.org/art/collection/search/336215 Public domain.[2]

I begin with a dream of my own that reflects the questions just posed:

Panic and Waking Up

In my dream I'm lying in my bed, happily curled up with a book when at one point I need to go to the bathroom. In my bathroom I notice that the faucet in the bathtub is leaking. I try to turn it off, but it doesn't work. The drainpipe is blocked and water is slowly filling up the tub, so I worry it might soon overflow and I send an email message to my plumber. I figure I'll go back to bed, but I feel uncomfortably warm so I go to turn down the thermostat, which works. But the hot air still keeps coming from the vents, and that surprises me. Maybe the plumbing people can fix that, too, I think. But before I could go back to bed, I smell smoke coming from the kitchen. A small fire is in the stove, so I grab the fire extinguisher and try to put it out. But it doesn't work. I get really worried now: maybe I should call the fire department and get out of the house. But when I reach the door, it doesn't open. I look to the windows but they're sealed shut, too. I need help and in the dream I feel panic. I start looking for my cell phone to call for help but I can't find the phone anywhere. Then, as panic becomes unbearable, I suddenly wake up.

As I wrote down this dream, I was reminded of the Four Horsemen of the climate apocalypse described earlier in this book. The Four Horsemen depicted by Dürer were death, famine, war, and pestilence. For us, in the climate apocalypse already underway, the Four Horsemen are different: drought, heat wave, fire, and flood—the threats discussed in earlier chapters of this book. We can speak of the prophetic Apocalypse, in the Book of Revelations, and approach it as a collective dreamscape—the "nightmare from which we are trying to wake up." Traditional theology would see the Apocalypse as a foretelling of future events. But it is already a parable of our present circumstance, and in this chapter I approach it as an exercise in "dreamreading."[3]

As in my dream, my "home" is planet Earth. In this dream the dreamer is threatened in multiple ways: flood and fire, but, above all, by being trapped. Dreams show us what we already know but cannot yet see.[4] We already know that there is no "Planet B." We are "trapped" on the earth. Jeff Bezos or Elon Musk may want to go to Mars. But we know that will not be the solution to the climate crisis. In my dream I can't go anywhere. I can't even open the door or get out through the windows. We're searching for safety, but we can't get out. I myself went on protest marches with young Sunrise advocates, but few answers came. We can call for help but how will we find the means to do so? Will collective panic become unbearable? Can we all somehow wake up?

It is a mistake to pose fear and hope as opposites, as if one is true and the other if false; as if one is desired and the other to be avoided. Fear and hope are of course a duality, a binary. But both are needed, not a false choice of one or the other.[5] Without fear any hope we have will prove to be false hope, a delusion, a wish fulfillment that is an escape from reality. Buddhist teacher

Pema Chodron[6] put it well: "Gloriousness and wretchedness need each other. One inspires us, the other softens us. They go together."

When we recognize the reality of the Unconscious we begin to understand that there is more in us than our waking awareness, more that we do not yet see. A dream like "Panic and Waking Up" is such a recognition. In that respect, it is a gift, a taste of fear and of the reality of our situation. Without such fear it is useless to look for suggestions of hope—in technology, in political action, in personal virtue. This is exactly why fear is necessary. Greta Thunberg put it best:

> I don't want you to be hopeful. I want you to panic. I want you to feel the fear I feel every day. I want you to act. I want you to act as you would in a crisis. I want you to act as if the house is on fire, because it is.

If "Hope is as verb with its sleeves rolled up" then the task is to roll up our sleeves. We won't do that without facing realistically the fear about our situation. "It's important to say what hope is not," says Rebecca Solnit in *Hope in the Dark*:

> It is not the belief that everything was, is, or will be fine . . . This is an extraordinary time full of vital, transformative movements that could not be foreseen. It's also a nightmarish time. Full engagement requires the ability to perceive both.[7]

Hope isn't the same as optimism. But it's different from defeatism, the voice in us that whispers "No, it's all hopeless".[8]

Elders, at some level in themselves, know that they—that is, we—are no longer young. They—that is, we—are closer to death than earlier in life. But the denial of death is very real, as Ernest Becker reminds us, even if denial is mostly unconscious.[9] Death is the limit of human time, and, as Heidegger understood, human existence itself (*Dasein*) is being-toward-death. We see here the parallel between aging and climate. Just as aging is the gradual encounter with the limits of our life, so climate change forces us, like or not, conscious or not, to recognize the limits of human enterprise on the earth. In "Panic and Waking Up" the dreamer comes across a series of these limits: the water doesn't drain out, the fire extinguisher doesn't work, the windows won't open, and the cell phone is lost. Every solution eludes the dreamer until panic ends the dream and the dreamer wakes up: "History is a nightmare from which I am trying to awake." Now "climate change is a nightmare from which we are all trying to awake."

In my work on dream interpretation, I tell people that dreams show us what we already know but cannot yet see. We could also say that a dream is its own interpretation—above all, in the feelings conveyed by the dream. In my dream "Panic and Waking Up," the key lies in the final feeling in the dream: panic becoming unbearable until the dreamer suddenly wakes up.

This dream, of course, is not just my private dream alone. It is what Jung would call a "big dream," a message for more than the dreamer alone. The dream comes at a moment of history when, more and more, we know what is already at hand and what will be coming soon enough: Here-Now-You-Hope. What to do? W.B. Yeats put it best: In dreams begin responsibilities. Greta Thunberg again: "Once we start to act, hope is everywhere. So instead of looking for hope, look for action. Then, and only then, hope will come." It is action that is called for, rolling up our sleeves so we can find genuine hope. But what action should that be?

I was recently asked to prepare some questions for a conference presentation on aging and climate change. The night before I was asked about this, I woke up with the following dream.

Running Out of Time

> I am at an event listening to a big speech about climate change. Afterward I made my way to the front to talk to the speaker, a man I didn't know. I told him it was the most impressive speech I'd heard in my life. I was at this event with a companion, a woman who hadn't made it up to the speaker. I had a message from her to give to him, something she'd written and signed. But it was signed not by her name but just a series of X's: x x x x x The speaker asked me about this but I couldn't explain it.
>
> My companion finally arrived and we were supposed to go to a restaurant afterward. She brought a child with her. The restaurant was too far so we went to one right across the street. I said it was getting late and we were running out of time that evening. I was worried because from where we were it would be hard to drive to my house in the time we had left.

Every character, every image, in this dream is me. And also it's you. We're all running out of time. From where we are now, it's going to be hard to drive home to solve the climate challenge in the time we have left. But we have no alternative except to try.

"The dream belongs to the dreamer" is what Velva Lee Heraty[10] said to me when both of us were co-leading a dream workshop at one of the national positive aging conferences I convened during my years working for AARP. In my own dream "Running Out of Time" I was listening to a speech about climate change from a speaker who I don't know—in fact that is the first uncertainty in this dream. Of course, I am the speaker, but perhaps I don't know myself either. My feeling in response was profound: the "most impressive speech I'd ever heard in my life." My female companion at the conference has an important message I am to hand off in the speech about climate change. But she doesn't give the message directly. Instead, I, as a dreamer, am the message-bearer. Like the speaker, she also doesn't reveal who she is: the message is "signed" with only a series of X's. The speaker demands to know what this means, but the dream only offers more uncertainty: "I couldn't explain it."

In the next part of the dream, my female companion and I are heading for a restaurant: we need more nourishment, including nourishment for the next generation, symbolized by the child who comes with us. So we go to the closest restaurant at hand, right across the street. But we're already running out of time. I am far from home, and I'm running out of time.

In Gestalt theory dream interpretation, every detail, every image of a dream is important, and every detail represents the dreamer: the dream belongs to the dreamer. "Running Out of Time" follows this pattern. It is filled with uncertainty, but feeling is always the guide to interpretation. The dreamer is overwhelmed by the importance of what he's heard about climate change. The feminine companion, what Jungians would call the anima, wants to communicate with a powerful message-bearer, the speaker. The feminine companion doesn't speak in her own name but is disguised. The speaker, who is the dreamer, doesn't understand the message and can't explain it.

When the dreamer and his companion finally get to the place of nourishment they can only reach a place nearby. Time does not permit going anywhere else. Future generations are present: the next generation, the child, is there in the dream, too, but the child never speaks. We can't hear the voices of future generations who will be most affected by climate change. The dream ends with the need for homecoming. The final feeling is one of worry: as with climate change, we are running out of time. Will we have time to act for the sake of future generations?

Dream worker Rachel G. Norment published an article "Let's Wake Up, Help Save Our World"[11] where she includes her own dream:

Stormy Weather

> I've gone to some building to get out of stormy weather. I find lots of other people there. Someone may question why I'm there. I say I've just taken refuge from the weather. I receive a phone call from someone I don't know. The person has heard of me and is begging me to warn other people about an impending disaster. I ask what she is talking about. The person says it's what was in the newspaper and it will happen in two days. I then remember seeing the item referred to. After getting off the phone, I turn to people near me and try to tell them what was said. No one wants to pay any attention.

Rachel Norment has worked as a facilitator with the Healing Power of Dreams Project, part of the International Association for the Study of Dreams, where I was a long-time member. In 2003, when Norment had the dream "Stormy Weather," the world had just experienced the cataclysmic effects of the Asian tsunami. By that date scientists were already documenting the melting of glaciers and other evidence of global warming. Only a few years later Al Gore would release his film "An Inconvenient Truth" and that film began to awaken larger numbers of people to what is invoked in "Stormy Weather." Norment's dream ends with the ominous feeling that "No one wants to pay any attention."

The dreamer, and others, would keep warning the world. For her, dreams were giving a clear message: "Wake up, everyone, before it's too late." As Joyce said: "History is a nightmare from which I am trying to awake."

Climate Nightmare?

> Christiana Figueres, age 67, is an international diplomat born in Costa Rica. She played a key leadership role in the successful 2015 Paris Agreement on Climate Change. But the Agreement did not yield all the results Figueres hoped for. For years, Christiana Figueres had a recurring nightmare. She saw seven pairs of children's eyes—the eyes of seven generations, staring back at her asking "What did you do?" Now, Figueres said, we have millions of children in the street, asking adults the same question: What are we doing?[12]

History ends now, the decade of the 2020s, even if history is not yet over. But right now, we are all the history from which we are trying to awake. In the remainder of this chapter, we will look at dreams at this particular moment of history, dreams that reflect both hope and fear about the climate crisis. James Baldwin famously said, "Not everything that is faced can be changed, but nothing can be changed until it is faced." So too, our dreams of hope and fear are one way in which people at this moment in history can face what needs to be changed. It is a mistake to see dreams as a flight from reality. Dreams show us what we already know but cannot yet see. Dreams are also a path to reality if we can receive the message: "In dreams begin responsibility." Dreams may lead us to a point where fear becomes unbearable, and we may find the means to wake up by facing the reality of our situation.

Hope and fear are often seen as alternatives, as being in opposition. In my dream "Panic and Waking Up" the dreamer moves step by step to discover that the problem, the plumbing problem, can't be solved. Then the kitchen fire becomes the threat. The final message of the dream is the feeling of panic itself. Where will hope be found?

One prominent climate activist has recognized the juxtaposition of hope and fear around climate change:

> Thanks to scientists and activists, the world has woken to the global existential threat of climate change, and anyone dedicating their life to reducing that crisis is making a necessary, rational, and heroic decision, but climate chaos is only part of the story. My fear (on less optimistic days) has been that, without a common and shared ecological view of our full array of planetary-scale impacts, those of us fighting environmental battles are often just shoveling sand into an incoming tide.[13]

Both fear and hope also belong to a deeper part of ourselves, a part not always known or recognized, which is why we call it the Unconscious. As one wit put it: The problem with the Unconscious is that it's unconscious. But dreams are a way into it. Dreams of fear about climate change show us what is already in us, at a level deeper than denial, deeper than everyday distraction. Normality and habit are a powerful pull, for people of all ages, but especially for elders. Gerontologist Robert Kastenbaum put it well when he said the best definition of aging is habituation.[14]

Facing a Future We Did Not Expect

Following Kastenbaum's definition, we might reply: elders today need to work against habituation in every possible way. I had the great privilege to know personally gerontologist Robert Kastenbaum, who described aging as habituation. Habituation means simply letting the past shape our expectations for the future. The problem is that we are living now at a time of VUCA: Volatility, Uncertainty, Complexity, Ambiguity. Habituation is a very poor position for facing a future we did not expect.

Those of us who are elders grew up at a time when we could take for granted democracy, free elections, and the rule of law. But today significant numbers of Americans don't agree with those principles.

"We're not doomed yet," says climate scientist Michael Mann on our chance to save human civilization. See Michael Mann's book *Our Fragile Moment*, How lessons from the earth's past can help us survive the climate crisis.[15] We want to believe, in fact, we do believe, that everything can go on just the same as it has always been. And it does go on the same as it's always been until an extreme weather event disrupts our world. Then we want to say, no, it can't be so, it's like a bad dream. The great spiritual teachers tell us that our ordinary daily life is itself a kind of dream. At certain moments—for instance, the death of a loved one—we may wake up. Such moments usually pass and then life goes back to normal. But an era of climate change means that extreme weather events have the potential to awaken in us feelings of hope or fear. Both feelings echo vulnerability, a realization that we are not in control. But we never have been in control of weather or climate, so our fear is entirely realistic.

The dream "Running Out of Time" brings us to a path of hope: Can we bring the message about climate change to others? The dreamer can't really explain the message: it's merely a list of x-x-x-x marks on the page. In the dream "Stormy Weather" the dreamer comes to the end where she feels "No one wants to pay attention." We fear being left alone with the message, left alone to face a climate disaster.

If you're reading this book, you, the reader, have the same fears, whether conscious or not. Perhaps you've received the message but then asked yourself "What can one person do?" That is the right question to ask. But it's only the beginning, only a question. It's stated the wrong way: one person. The answer is that one person alone cannot solve the problem of climate change. That's not the path to genuine hope. The path to hope is through action, the right action, which is not to be one person acting alone. One person alone cannot solve anything. But acting together, we may find ways that bring hope. Greta Thunberg puts it this way: "Once we start to act, hope is everywhere. So instead of looking for hope, look for action." Remember always, David Orr's words: hope is a verb with its sleeves rolled up. It is not optimism, it is not wish fulfillment, and it is not waiting or wanting something good to happen, for a solution to emerge. As Terry Eagleton put it, there can be hope without optimism.[16] On climate change, we need the right kind of optimism.[17]

Twelfth-century Jewish philosopher Maimonides said that hope is belief in the plausibility of the possible, as opposed to the necessity of the probable. More recently, German philosopher Otto Friedrich Bollnow saw hope as the touchstone of human emotion and existence: "Hope thus points to the deeper ground in which the feelings of patience and security are rooted, and without which [we] would never be able to relax [our] attention or go to sleep tranquilly." In sum, hope is what "comes to us without any effort on [our] part, as a sort of gift or grace." Hope is a frame of mind that connects us to the future, not as the inevitability of our own death, but as "an infinite source of new possibilities." Because the future contains these infinite possibilities, we are not paralyzed but energized.[18] Hope is the belief that the future can be brighter and better than the past and you have a role to play in making it better.[19] HERE NOW YOU HOPE

In short, hope is not passivity, which is what pessimism and cynicism encourage. In our own era, negative voices try always to be sophisticated by showing all the ways in which efforts can fail. Many of us, elders and young people, fear that the future, including climate change, will be a disaster. What we're seeing already may not quite be "deaths from despair." But it could prompt passivity and paralysis: "What's the point?" Doomscrolling is the biggest enemy faced by climate advocacy. As faith in the future plummets and the present blends with the past, we doomscroll and catastrophize and feel certain that we've reached the point where history has fallen apart.[20]

The Jewish people in the time of the Roman Empire had every reason to feel that history had fallen apart and that there was no future. The reaction of the wisest was different:

> The great first-century rabbi Yochanan ben Zakkai taught that "If you have a sapling in your hand, and someone says to you that the Messiah has come, stay and complete the planting, and then go to greet the Messiah.". . . A hadith of the Prophet Muhammad says, "When Doomsday comes, if someone has a palm shoot in his hand he should plant it."[21]

Even when we feel that history has fallen apart, we must find hope in a very different direction—neither through certainty nor rationality but in hope that comes from dreams. As the voice of God says in the Bible, "When the last days come, I will pour out my Spirit on all people. Your sons and daughters will prophesy. Your young will see visions, and your old ones will dream dreams" (Acts 2:17). In short, we need all generations, both young and old, working together. They will find hope in different ways. The way of prophesy and visions of the future belong to the virtues of the young. Dreams belong to the virtues of the old.

There are dreams that reflect premonitions of environmental disaster, and we need these dreams. Here is a dream from Paco Mitchell:[22]

Diamonds and Tornado

> I am in a house with several others. A tornado is coming. We prepare for it by practicing sky-diving maneuvers—ways to stay in touch as we hurtle around inside the vortex of the great whirlwind. There will be no escaping the tornado. In fact, a square hole has even been built into the ceiling of the room for the explicit purpose of permitting our absorption into the massive tornado. The last thing we have to do before the tornado hits is to swallow a handful of diamonds.
>
> When the tornado finally arrives, the atmospheric pressure drops and we are all sucked up in the turbulence. As we whirl around with the debris inside the giant funnel we try to execute our 'maneuvers' to stay in touch. The experience is awesome and frightening, but when I remember the diamonds I have ingested, I know that—whenever and wherever I land—the diamonds will be with me and will form the basis of a new life.

Paco Mitchell understood his dream to furnish him with a personal orientation and guidance, a form of wisdom, as he says, during a time of upheaval and change. He notes that "Diamonds and Tornado" should be read as a collective dream, since we are all facing a tornado of planetary proportions. Books such as *Dreaming the Future* (2012) by Kenny Ausubel, co-founder of the Bioneers Network, point to threats around us likely to grow larger in decades to come: global warming, peak oil production, diminishing energy resources, and worldwide financial turbulence.

Paco Mitchell's dream, he believes, is a message for us all to "stay in touch," as in the dream "Stormy Weather." In Mitchell's dream the dreamer is urged to do so by practicing "maneuvers." We cannot stop or even control the planetary forces gathering strength. But the dream suggests another message. Along with "staying in touch," each of us must do something else: we must "swallow a handful of diamonds." What could be the meaning of this image? Mitchell believes that diamonds represent essential values that "must be incorporated, assimilated, embodied." He alludes to the "diamond body" of Buddhism, reminding us of the connection between spiritual insight and environmental action. "Diamonds and Tornado" is a collective dream message, and a warning to us all.

In his preface to the 2006 edition of *Man's Search for Meaning*, Rabbi Harold Kushner goes to the heart of Viktor Frankl's contribution by saying: "Forces beyond your control can take away everything you possess except one thing, your freedom to choose how you will respond to the situation."[23]

11.1   Viktor Frankl: Eco-Elder

Figure 11.2 Viktor Frankl. Public domain.

Only once did I see in person Dr. Viktor Frankl, author of the celebrated book *Man's Search for Meaning*. But long before I saw him speak in person his book had influenced me from the time I was in college (1963). I have never forgotten the experience of seeing Frankl in person.

Theology professor Melvin Kimble brought Viktor Frankl to the annual conference of the American Society on Aging, where Frankl spoke to a crowd of over a thousand listeners. All of us were riveted to our seats as we watched this little man on the stage, displaying the extraordinary power of "late freedom." I had seen that same power when I was present to see a very aged and frail Arthur Fleming addressing the plenary session of the 1995 White House Conference on Aging. Fleming was appointed Secretary of Health, Education and Welfare by President Eisenhower. In later years Fleming became director of the U.S. Administration on Aging. When Fleming spoke, as when Frankl spoke, cases, one could hear a pin drop as we all waited, not in vain, for words of wisdom in old age. We were disappointed neither by Fleming nor by Frankl.

Frankl's message gives a deeper way to approach hope and fear about climate change. As a psychiatrist, he knew how we need to approach both hope and fear in the deepest part of our psyche. Only in that way can we find the freedom the act and to survive, as Frankl did in desperate circumstances.

In Frankl's case his message was born from the experience of surviving four different concentration camps, including Auschwitz, while others around him died or became hopeless. In our current climate crisis it is essential to remember Viktor Frankl's admonition about our freedom to respond. In the case of climate advocacy, "Our reaction to the situation depends on whether we find a way to see our lives as still meaningful."[24]

We cannot move toward genuine hope without first acknowledging the reality of the threat, which is why fear is needed, too. As in other situations, climate emotions demand both sides[25]—here the "reality principle" (Freud's term) for maturity. Threats are the Four Horsemen of climate apocalypse: fire, flood, drought, and heat waves. Some of these more easily appear as dream images. In the remainder of this chapter, we explore dreams of fear and hope, especially those where fire and flood are images that convey the message of the dream.

In an earlier discussion, we looked at the Four Horsemen of the Climate Apocalypse: flood, wildfire, drought, and heat wave. These threats provoked the impulse to search for safety, and that is the leitmotif of the following dream, "Feeling of Panic."

Feeling of Panic

> I dreamt my family and I were trying to find a place to live. Tennessee, where we are now, was no longer safe. Volatile weather meant you could never make plans: tornadoes and floods and lightning and straightline winds. But every time we had an idea where we could live that was safer, that place would have a disaster. There were fires, floods, and avalanches. The sense of futility, hopelessness, and panic was palpable.[26]

The dreamer understands that you can never make plans; you are no longer safe in her home state of Tennessee. But in the dream searching for safety does not work: every place found turns out to be a disaster: fire and flood followed by hopelessness and panic.

Both hope and fear can be present in the same dream, as in this dream of my own, "Fixing the Computer."

Fixing the Computer

> I was with an unknown person who was helping me with my computer. At one point he asked me, Why do you care so much about climate change? I answered, Because I care about the people who are coming here after us. He replied: But you don't even know them—you have no relationship with them. I immediately answered, but I **DO** know them. They're my children and my grandchildren and many, many others. Then we started working on the computer again. But nothing he did fixed the problem. I looked at the screen and saw images of people. But I couldn't hear their voices.

In my dream the computer is the symbol of the world machine or megamachine as described in *The Myth of the Machine* published by Lewis Mumford in 1970, just a few years before I met him. In my dream I am trying to fix the world machine that is destroying our planet with climate change. The unknown man in the dream asks why I am so concerned about climate change and he objects to that concern because we can't even know future generations. I respond by citing my children and grandchildren, those who I know. When he try to fix the computer, the world machine, I can see faces, the image of future generations, but I cannot hear their voices. I am separated, concerned but helpless.

Impending Doom

I was in a group of several young women (and was younger than I am now) traveling in a boat without a rigid hull that kept shifting shape and taking on water. I had a camera with a telephoto lens that I was having a lot of trouble keeping dry.

We stopped at a small island, and were diving in shallow water just off the shoreline near a pier for starfish and conch. As the tide started to go out, we thought we'd be able to gather more from the now revealed sea bottom, but we found everything exposed was dead and decaying . . .

. . . You could see the islands had been reduced to piles of sand because of storms and development. You could see how the islands had been dredged and reshaped to create "driveways" for boats to reach the houses of the wealthy. But now the houses were abandoned and the waterways among them were filled with stagnant brown water.

One of the young women asked why the water was brown now instead of the clear blue in which we had been diving. I answered (back to my current age) because they are full of shit, from the boats emptying their waste directly into the water.

We flew over island after island, silently, bearing witness to the devastation. The people were gone. There was no greenery. The waters were stagnant between the various islands. Eventually, the plane ascended, and we flew away feeling a sense of impending doom.[27]

The ties between generations, between young and old, are been torn up by the loss of confidence in the future. This level of fear can be expressed by images of horror: for example, the idea of a Zombie, or a reanimated corpse, depicted famously in the film *The Night of the Living Dead*, which came out (1968) when I was myself the same age as Daniel Sherrell was when he had the dream recounted in his book *Warmth: Coming of Age at the End of Our World*. Here is another dream where the rupture of family ties is conveyed by the image of zombies:

Zombies in the Family

It was a horrible dream . . . I had lost contact with my family. I had a last phone call and then there had been a cut off . . . then months later somebody found a phone that had belonged to one of my parents . . . and there were photos in there of my [family]. I woke up with this horrible . . . sense of anguish, of they were dead and gone . . . And it stayed, like that for days . . . Which after having made light—"let's talk about zombies next time"—wow it's quite something isn't it?. I think the initial meaning of it is pretty straightforward; it made me feel if there are dangers in the world, it's the people we love the most . . . that's how we feel it right? We see it as death of the family.[28]

When I was ten years old (1955), The Family of Man exhibit opened at the Museum of Modern Art in New York City, not far from where I was living. It was hailed at the time as the most successful exhibition of photography ever assembled. The optimism about the universal human is clouded today. Can we speak about a global family at a time when 8 billion people all face different threats from the climate crisis? As the dreamer in "Zombies in the Family" puts it, "there are dangers in the world" and the level of fear in this dream can be compared only to a death in the family, even including the Family of Man.

In *Braiding Sweetgrass*, Robin Wall Kimmerer writes:

> Joanna Macy writes that until we can grieve for our planet we cannot love it—grieving is a sign of spiritual health. But it is not enough to weep for our lost landscapes; we have to put our hands in the earth to make ourselves whole again. Even a wounded world is feeding us. Even a wounded world holds us, giving us moments of wonder and joy. I choose joy over despair.

11.2   Joanna Macy: Eco-Elder

You live inside us, beings of the future.
In the spiral ribbons of our cells, you are here . . .
You who come after, help us remember: we are your ancestors.
Fill us with gladness for the work that must be done.[29]

Joanna Macy's environmental advocacy occupies a unique niche in the domains of science, policy, psychology, and technology. Before each of us can respond to these different domains, we need to appreciate how we experience life around us emotionally and to see how I we manage the emotions attached to our thoughts.

This practice of self-awareness is the genius of Joanna Macy in *The Work That Reconnects*, which came out of Macy's academic work on Buddhism and Complex Systems. Her perspective became translated into both practices for personal development and workshop rituals that reinforce social support. Kathleen Schomaker, Director of Gray Is Green, the National Senior Conservation Corp, came to appreciate Joanna Macy's work in very personal terms. As Schomaker puts it: "When I found myself mired in grief, despair and cynicism, Joanna's guidance, teachings and group processes were my way through." Schomaker stresses that Macy's approach does not promise a "way out." Instead, Macy teaches process skills of "going through," thus avoiding the pitfall of pretending we to emerge from despair and cynicism.

To engage with Joanna Macy's work is to take up gratitude—to start there, over and over again. As Kathleen Schomaker puts it: We begin to notice how grief and despair arise from witnessing what we have done to that for which I feel deep gratitude—air, waters, plants, soils, other species, and other humans. In noticing, if we pay close attention, we may see how grief and despair reveal deep love and reverence for that world now suffering from many wounds inflicted by human beings.

Joanna Macy's *Work That Reconnects* leads us to a positive question: How can I not be devoted to caring for that which I love so deeply? It is essential not to be overwhelmed or to become cynical. Instead, we hope to grow in awareness of the human communities devoted to deeper ways of understanding and healing. To follow Joanna Macy's path can lead to a "Going Forth" with an awakened heart and mind. This shift in awareness can send us out into the world of change, where each of us acts from a small corner of the planet but with the joy and peace needed to stay with the practical work each has been given.[30]

Robin Wall Kimmerer again:

"The trees act not as individuals, but somehow as a collective. Exactly how they do this, we don't yet know. But what we see is the power of unity. What happens to one happens to us all. We can starve together or feast together . . . The moral covenant of reciprocity calls us to honor our responsibilities for all we have been given, for all that we have taken. It's our turn now. Whatever our gift, we are called to give it and to dance for the renewal of the world."[31]

Acts of restoration are what it means to have acts of hope:

We need acts of restoration, not only for polluted waters and degraded lands, but also for our relationship to the world. We need to restore honor to the way we live, so that when we walk through the world we don't have to avert our eyes with shame, so that we can hold our heads up high and receive the respectful acknowledgment of the rest of the earth's beings.

Dreams of Hope

The idea of "hope" itself is ambiguous, even ambivalent: is it an indicator of a positive future or is it just another form of wish fulfillment, even escapism? As usual, the poets have found the most concise way to encompass the different faces of hope. For example, Emily Dickinson writes:

"Hope" is the thing with feathers—
That perches in the soul—

And sings the tune without the words—
And never stops—at all—

Hope is like the bird that never stops singing, even if we can't know the words of the tune being sung.

Many American Indian tribes understood dreams to be meaningful, messages from the spirit world revealing an order to the universe usually hidden from everyday conscious life. They recognized that dreams could be a source of guidance about whether wishes or hopes should be gratified. Psychoanalytic philosopher Jonathan Lear[32] gives the example of a dream by Plenty Coups, a leader of the Crow people in the 19th century. Plenty Coups was given instruction in his dream by a man telling him as a very old man sitting in the shade of a tree: "I felt pity for him because he was so old and feeble. 'Do you know him, Plenty Coups?' he asked me. 'No,' I said . . . 'This old man is yourself, Plenty Coups." This dream is the dream of an old man recounting a dream remembered from when he was a boy. But the dream was told to the entire tribe and soon become incorporated into the tribe's own self-understanding.

Some would dismiss this dream as wish fulfillment by a Native America tribe under stress by the encroaching white man's culture. But Jonathan Lear sees it as an example of what he terms radical hope, "a manifestation of imaginative excellence." "It enabled the tribe to face its future courageously—and imaginatively—at a time when the traditional understanding of courage was becoming unlivable."[33] The radical hope called for by Native Americans in their circumstances is Lear's account of ethics in the face of cultural devastation. Radical hope applies very much to our condition today facing the devastation of climate change.

Here is a climate dream where the message from the "beyond," from another dimension, conveys to the dreamer a message of hope:

We Can Heal Anything

> My friend and I are leading a team that is trying to save the planet. Despite our best efforts, we discover that full planetary extinction is imminent. A feeling of profound grief brings me to the ground. At which point, two human-sized entities appear, seemingly from the stars or another dimension. With a supreme presence of pure love connecting us, they telepathically convey the following message:
>
> Forget everything you think you know.
> We can heal anything in an instant.
>
> Then I am teleported to a large auditorium, in which I repeat the message to hundreds of people.[34]

"We Can Heal Anything" is important dream about hope because it begins at the opposite extreme: with feelings of fear and grief, a grief that brings the dreamer "to the ground." The mysterious entities coming from the stars or from another dimension are human-sized: in other words, available to our human world, not entirely supernatural. The feeling conveyed these transcendent forces is one of "pure love" and the connection is telepathic, not by ordinary methods. The message in the dream tells us to forget everything we know and then, by the end of the dream, the dreamer stands in front of a crowd in an auditorium, repeating this message of hope to multitudes of people.

Is delivering the message enough? Is writing a book about climate change enough? I have often quoted David Orr's statement about hope: "Hope is a verb with its sleeves rolled up." In the following dream the dreamer anticipates action on climate change. As in all interpretations of dreams, the key to interpretation comes from the feeling conveyed to the dreamer:

Hope for Change

> I dreamed we had a summit of regular people to decide how to push for action on climate change. One path involved talking to and influencing diplomats somehow. We all felt hopeful by the time I woke up.[35]

This dream begins with collective action—a summit meeting, but one that involves "regular people," like the dreamer, like you and me. The task of the summit is practical: how to push for action. The target audience for action is beyond regular people. The path is "influencing diplomats," people in positions of power or influence, even if the method is not yet clear ("somehow").

Dreams show us what we already know, and we may even be conscious of it: for example, as in feelings of hopelessness. In a dream it is possible for the dreamer to confront the feeling of complete hopelessness about the climate crisis, as in this case when the dreamer wakes up and feels hopeful. It sounds like a paradox, but it has rightly been said that a dream is its own interpretation. The key lies in the feeling conveyed: "We all felt hopeful by the time I woke up."

We live in a time when people think of themselves are more sophisticated and in touch with reality if they are pessimistic, even cynical. Thomas Merton, the mid-20th century monk who helped to recover the roots of spirituality, understood that it was easy to lose hope or be uncertain how best to act. That is why he insisted that we do not need to know precisely what is happening, and this is true above all for climate change. What we need is to recognize the challenges of the present moment. That's what this book has been about.

One response of both elders and young people is to think "I didn't know it was this bad." What some have called "The Knowledge" about climate change can be devastating:

> How do you respond to your encounter with "The Knowledge?" How do you react to the terrifying realization that our biosphere and our

future are in grave and imminent danger? Do you retreat into the comfort of denial, hopium, and inactivism, or do you take a stand for the planet and the future?[36]

Is it time for "Reality Therapy" about climate change? The following dream, "All The Truth," is one where the dreamer herself, when fully awake, took what she sees to be "pretty much all the truth."

All the Truth

[I] had a dream that I was watching a horrendous news report on wildlife trapping, hunting and extinction and legit woke up crying. Worst part is, I'm fully awake now and it's pretty much all the truth.[37]

In "All The Truth" the dreamer afterward claims to be "fully awake." But that sentiment is limited by whatever degree of consciousness the dreamer has. Those limits include prejudices, assumptions, and self-limiting beliefs, which may not be conscious at all. "The problem with the Unconscious is that it is not conscious." The dream "All The Truth" reminds us about interpretive beliefs embedded in what we think of as our conscious mind.

The need to confront self-limiting beliefs of all kinds means we need always to confront the reality of repression, of denial within ourselves. The following dream appears on the last pages of Per Espen Stoknes' important book *What We Think About When We Try Not to Think About Global Warming*.[38]

Flowing With the Winds of Change

Twenty years ago I had a dream that has stuck with me since. I was heading up some snowcapped mountains on a snowy road to spend time at a cabin on the other side. But I couldn't get there. A friend had given me a keycard to the sky lift that could take me there, but I had left it down below, in my car, and now I was somehow in a bus, but the bus had gone too far. Off the bus and walking back, I met Picasso along the road. He showed me a painting of a simple Picasso-style tree bending in strong winds. "It's titled Activist," he said.

Stronger winds are blowing now, and not only the hurricanes . . . The disruptions of weather and air are also driving societal, cultural and inner shifts. Some long for a cabin outside it all, beyond the hills in undisturbed nature. In the dream, this move to get out of all is frustrated. When I try to opt out with ski lifts, cars, and buses, they bring me back to the road. Where surprisingly Picasso—the great imagist—is waking. His picture shows me as an Activist, he says. But the real activist is not the tree that is pictured. It's the invisible wind who is the activator, bringing the tree into movement."

Hope is tied to future generations, just as fear is linked to future generations. In Stoknes's dream, it is possible for the dreamer to confront the feeling of complete hopelessness about our environmental situation but also feel the opposite. In this dream, as in others, a key to interpretation comes with an extended version of the Gestalt principle that every person, every detail in the dream somehow represents the dreamer:

Having Hope

[There was] an old friend . . . who was very caught up in this view of the world around climate change which was that it was all completely and utterly hopeless . . . I was sort of talking at her . . . haranguing her . . . "The children," I said, "You can't raise children without any hope". . . that just kind of horrified me.[39]

Dreams can point us to action in the future. For example, Elias Howe invented the sewing machine because of dream he had. In Howe's dream he encountered cannibals surrounding him and when he woke up he remembered their spears, which had holes in the them, moving up and down. The image of the dream inspired his design of the sewing machine.

The following dream is remarkable not only because of its link to future generations but because the dreamer himself, Ward Greene,[40] awoke from the dream with a very specific plan of action, including even the very name of the course of action he felt guided to pursue:

Caring for Future Generations

I woke up with a start. My heart was racing—I knew I was late. I must have forgotten to set the alarm. I needed to get to the SAGE meeting. There were going to be hundreds of seniors, folks like me, who were committed to helping the next generation. They would be waiting for me.

I started getting dressed before I realized it was all just a dream. I told my wife how real it felt, how anxious I was. She laughed, reminding me that I never have first person, realistic dreams. She then went and got a copy of a literary magazine; she wanted to share something she had just read. She said it sounded like my dream. It was a Greek proverb:

"A society grows great when its elders plant trees whose shade they know they shall never sit in."

Hearing that proverb took my breath away. I nearly wept. Although there was no such organization as SAGE, Senior Advocates for Generational Equity, that name was as clear in my dream as if I had seen it on a banner.

This dream came from Ward Greene, who, inspired by his dream, went on to create Senior Advocates for Generational Equity (SAGE) as an actual thriving enterprise, which it has been now for over a decade. His dream conveyed to him the need for generations to work together on environmental justice[41]

Ward Greene's dream was so powerful that when he woke up the message of the dream completely took over waking consciousness. As Yeats put it, "In dreams begin responsibilities." Greene was fortunate that his wife had just read the Greek proverb about elders planting trees for future generations. An example of synchronicity? A fortunate coincidence? It doesn't matter. What matters is the feeling that now awakened dreamer had, a feeling that aspiration to create something that did not exist. And it happened. Ward Green created SAGE as a thriving enterprise.

We hear repeatedly the Amazon forest is the world's biggest repository of vegetation that helps in our struggle with climate change. The following dream is from Rob de Laet, a Dutch protector of the Amazon rainforest. De Laet divides his time between the Netherlands and Brazil, where he lives in a valley in the rainforest and works closely with the nonprofit advocacy group Climate Change & Consciousness and with local residents, including indigenous peoples, who appreciate the importance of preserving the Amazon rainforest.

Hope From the Amazon

In my dream there was a huge procession of shamans and chiefs from indigenous peoples of the Amazon in the rain forest. There was huge ceremony of mourning for all that had happened to themselves and their environment and then there was forgiveness. Then that space opened up and there streamed in all the climate youths, a lot of organizations and companies ready to regenerate the world, the forest, society, themselves. And there was a rippling of hope going from the

Figure 11.3 Amazon River in the state of Pará, Brazil. Getty Images (Anderson Coelho)

Amazon rain forest over the planet. When I woke up I said to myself, the forest has given me a job so I had better start getting to work. I realized later that while we're in a complete and utter mess there's still a lot of hope.[42]

The Amazon rainforest is a vast assembly of trees and vegetation of all kinds. But can we see the drama of climate change in a single tree? Some of us can, including noted dreamworker Steven Aizenstat.[43] Here is his dream.

An Old Oak Tree

I dreamed about an old oak tree, one that had lived on my property for well over 70 years. That old oak tree unexpectedly got uprooted in a strong wind-storm. That same night, the oak tree appeared as an image in my dream.

About this dream Steven Aizenstat said:
What is this tree expressing about itself through the dream image? What is the "spirit" in the tree asking on behalf of itself?. . . Death of the oak does not occur overnight. The natural process of decomposition takes place over time, life gradually leaving the leaves, limbs, trunk, and roots. Dream images which reflect this process of dying may continue to present themselves to the dreamer as the felled oak in the world continues to dream.

Jonas Salk, alluding to this challenge of working on behalf of future generations, put our problem well when he said, "We are being bad ancestors." How then can we build what is required for generations to come? How do we become what Roman Krznaric called "Good Ancestors?"[44]

Eco-philosopher Joanna Macy has described the challenge of our time as "The Great Turning." A clue for how to think about the challenge is provided in a dream recorded by a young physician, Max Zeller, in the year 1949, when Zeller met with Carl Jung. Jung spoke to Zeller before the young physician left Europe for the USA. At that meeting, Zeller reported the following dream:

Building the Temple

A temple of vast dimensions is in the process of being built. As far as I can see there are incredible numbers of people building on gigantic pillars. I, too, was building on a pillar. The whole building was in its first beginnings, but the foundation was already there. The rest of the building was starting to go up, and I and many others were working on it.[45]

Jung commented on Zeller's dream by identifying the "temple" here as the construction of a new world culture, a transcendent platform beyond any single individual. Like Zeller in his dream, each of us works on a single pillar, but "many others were working on it."

Following the dictum of Montague Ullman and Jeremy Taylor,[46] "If this were my dream," I have my own vision of building the temple. In my vision dictum we are working on this temple of vast dimensions myself. It is a "temple" for climate advocacy, for preserving our precious earth for future generations. As in Zeller's dream, I am building on a pillar, and the foundation is already there. I see myself as one person in a vast crowd of people who are working to build the structure. In my vision I see some who I've met in person, people like Jane Goodall, Rabbi Zalman Schachter, Alexander King, and Lewis Mumford. I see others whose faces I can recognize like Wendell Berry, Joanna Macy, James Hansen, Christiana Figueres, Greta Thunberg, and David Attenborough. Some of the workers in the building are famous, but thousands of others are unknown. All are working strenuously as the rest of the building is starting to go up, and we are all working on it together. No one person alone could do it. But together, we have hope of success.

Here we touch on the duality between individuality action and collective action. Collective action, and the symbol of hope, is represented in this dream as a "temple of vast dimensions". Let me offer another dream that presents a powerful image of this link between individual effort and collective action, which means leaving a legacy for those who come after us in climate advocacy.

A Difficult Task

> I have been set a task nearly too difficult for me. A log of hard and heavy wood lies covered in the forest. I must uncover it, saw or hew from it a circular piece, and then carve through the piece a design. The result is to be preserved at all cost, as representing something no longer recurring and in danger of being lost. At the same time, a tape recording is to be made describing in detail what it is, what it represents, its whole meaning. At the end, the thing itself and the tape are to be given to the public library. Someone says that only the library will know how to prevent the tape from deteriorating within five years.[47]

This dream begins with a recognition of the enormous scale of the climate crisis: "I have been set a task nearly too difficult for me," and also the dreamer's fear that he may not be up to a task which is "nearly too difficult" for him. But the dreamer is not completely alone. The dreams need to go into the forest to find the covered log and uncover it. Then comes a struggle in carving a design "to be preserved at all cost," a precious accomplishment, like finding "the Pearl of Great Price" mentioned in the Gospel of Matthew. But help comes in the message of the tape recording, which must describe "in detail what it is, what it represents, its whole meaning." Finally, there is the dreamer's obligation to give the "thing itself" and the tape recording to the library, a repository of wisdom for past and future generations.

The first task is active, something we must do: to find the covered log in the forest and carve the design, a work of craftsmanship. The second task is more reflective: to make a tape recording, a commentary about the whole meaning of what the dreamer has done. These two sides of the challenge represent two sides of our common human journey: acting and reflecting. There is also a duality in the setting of the dream. It begins in a dark forest but concludes in a library, a symbol of higher culture. The challenge cannot be completed unless we bring together these two sides of ourselves, forest-dwelling and culture, acting and reflecting. To bring these two together is "a difficult task" which requires courage to enter the dark wood, just as Dante needs to do at the beginning of the Divine Comedy.

These stages of the soul are represented in a modern cinematic myth *The Wizard of Oz*. That film, and the book, tells the story of Dorothy's journey, after a Tornado, from her hometown Kansas to the Land of Oz. Her story is a quest for homecoming, as the Persian poet Jalal ad-din Rumi evokes homesickness in the opening lines of his masterpiece, the Mathnawi. When Dorothy wakes up in Oz, she knows she must return home, but cannot find the way. She is guided by three figures, the Tin Man, the Lion, and the Scarecrow, each representing deficiencies that need to be overcome to reach a fully human state. Dorothy's struggle is with the Wicked Witch of the East, and her breakthrough comes when she realizes that the Wizard of Oz himself has no power to help her complete her quest. She had the power in herself all the time, just by clicking her heels together. Her Return

Figure 11.4 The Wizard of Oz: Dorothy's helpers on the journey. Public domain.

to Kansas represents her own growing up and coming of age, just as the spiritual quest, the five stages of the soul, reflects the sequence in which we "come home" to where we have always been.[48]

Let us come back now to another famous dream, this one the dream of Jacob in the Hebrew Bible. The central image of that dream, "Jacob's Ladder," was understood by commentators, Jewish, Christian, and Islamic, as a symbol of the channel of revelation between the sleeping prophet and the unseen world of the spirit. Jacob's dream was a breakthrough experience of the highest order. Upon awakening, Jacob proceeded to "make sacred" the place where he had been sleeping. He anointed the stone where his head was resting, accepting Yahweh on behalf of himself and his descendants, thus leaving a legacy. He brought back his gift as a boon to those who would come after him: "In dreams begin responsibilities."

Jacob's son, the Prophet Joseph, continued this legacy work begun by Jacob. Joseph is the first dreamworker mentioned in the Bible. He interpreted the dreams of Pharaoh and then went on to become a counselor to Pharaoh and acted wisely to prepare Egypt for the coming famine. In short, Joseph did more than interpret dreams or even find his own salvation by being released from prison. He went on to sustain the salvation of the whole of the people, which is the challenge of "healing the world" (*Tikkun Alam*) and leaving a legacy.

Consideration of dreams and climate leads us finally to appreciate the importance of dreaming, not only for individual psychology or self-realization but for the wider society. As the psychiatrist Anthony Stevens has noted:[49] "To work on dreams, therefore, is not a petty form of self-indulgence, but a spiritual ritual of cultural and ecological significance: the more conscious we become as individuals, the more hope there is for our tiny portion of the universe." And hope is what must never be lost.

There is a deep existential and spiritual dimension of hope. In this book, I have emphasized the Four Horsemen of the Climate Apocalypse and also the ethical demands on people of all ages, young and old, to work against these forces. At the same time, our dreams remind us that there is a greater dimension in depth, both to ourselves and to the world around us. This dimension of depth, which I expounded in my book *The Five Stages of the Soul*, is also essential to avoid the horizontal flatness of some environmental ethics. As Reza Shah-Kazemi has put it:

A secular set of recommendations regarding environmental ethics is flat and horizontal, lacking the dimensions of height and depth which are bestowed by a sacred view of the environment, one in which wholesome attitudes towards the natural world are rooted in the principle of [Unity], which engages all levels of the "environment," from the material to the moral to the spiritual and the Divine.[50]

For those who look deeply into themselves and into the cosmos, in Shah-Kazemi's words, there is no justification for despair or despondency "because the more one sees the signs of the impending end of this world, the more one is urged to do whatever is possible within the sphere of one's own competence, within one's own 'world'. This is based on the principle of man as a microcosm, a 'little world'; the whole of the world is, in some way, affected by the individual, for the individual is a recapitulation of the world, a reflection of the world; and the reflection cannot ultimately be separated from that of which it is a reflection." As Rebecca Solnit[51] and others have repeated: It is not too late: "The more clearly you see the 'signs' of the end of the world, the more resolutely you must train your focus on what you, individually, must do about your world . . ." The vertical dimension calls us to hope, which "is a verb with its sleeves rolled up," in David Orr's words.

11.3   Jan Hively: Eco-Elder

Figure 11.5 Jan Hively. Used with permission of Jan Hively.

I met Jan Hively at an early Positive Aging Conference, one of a series of conferences I convened under sponsorship by AARP. Little did I know that Jan Hively would go on to create an ongoing global network of positive aging connections in nearly 70 countries around the world. This Pass It On Network, co-led by Moira Allan in Paris, is based on the principle that older people are called to "pass it on"—that is, to share know-how with others and respond to common problems such as climate change. Only now are scholars recognizing the power of these networks to change the world.[52] Jan Hively recognized it long ago.

Jan Hively was born in 1932, and she did remarkable things earlier in her life. But she has said that the best years of her life were her 70s. That's when she co-founded three grassroots networks connecting positive aging advocates. From the time I met her, she has always been for me the model of lifelong learning I have aspired to. Jan said, over and over again "I believe that we learn to be elders in later life." She has been for me the model of what an elder could be and should be. The key element is her commitment to work on shaping the future for generations to come.

In her later life, I could recognize familiar figures who had attracted me. In her mid-life, Jan met Maggie Kuhn, founder of the Gray Panthers, whom I met at the first Conscious Aging Conference in 1992. Jan embraced what Viktor Frankl said in *Man's Search for Meaning*: All we really have control over is how we perceive our experience and how we think about our lives. As a scholar of comparative religion and a student of dreams, I particularly appreciated the path Jan felt called upon through a dream she had that conveyed this message: "Seek the Everlasting."

Jan Hively is much more than a dreamer or a talker. She served as Deputy Mayor of Minneapolis, and she was named a Purpose Prize Fellow. She works tirelessly to connect people and facilitate small discussion groups with people who share deeply and collaborate on joint efforts for the common good. She is working on the issue of greatest concern to us all: climate change in an aging society. She gives the next generation hope as she has given me hope.[53] W.B. Yeats put it best: "In dreams begin responsibilities."

Here is Jan Hively's dream:

Seek the Everlasting

In the dream, I was looking through a book of prints and paintings. The book was large, the size of a wallpaper sample book. There were some prints and some acrylic paintings that were wonderful. I tried to decipher the signature of the artist as I was looking at a painting of clouds. Then suddenly the painting opened up and I stepped into it and walked into the clouds. From behind my left shoulder, I heard a deep voice say:
    "Seek the Everlasting."
    I felt overwhelmed and stayed partially awake through the rest of the night.

Jan Hively added: "That dream has stayed in my mind and been really significant for me since I experienced it in 2009. I see us human beings as one of millions of species who have evolved on this planet. And our planet Earth is just one of the millions of stars and planets in the universe. In the midst of our climate crisis, the most important thing we can do is to nurture and protect the capacity of the earth to heal and renew itself as part of an ever-lasting universe.

> While pondering the aftereffects of last week's dream webinar, the light dawned for me, and I felt overwhelmed with gratitude. Each of my two children has given me an extraordinary gift by contributing directly to the capacity of the earth to heal and renew its place in an everlasting universe. My daughter is an environmental engineer, and my son is a soil and water scientist. What greater gift to the everlasting could I give birth to in my life children who are actively contributing to the health of our universe?[54]"

Hope and Fear Together

We need them both. It's a false choice to pick one or the other. Are we ready for the climate apocalypse?[55] Is it true that if only people knew how bad things are, they would act to prevent it? Yes, let's take in all the facts, no matter how bad it seems. But we need to remember the story of Cassandra and her warnings to the city of Troy that it would be destroyed. That didn't work for Cassandra, did it? As with Cassandra, more warnings seem to have left us short of our climate destination.

What would it take for warnings to be heard?[56] First is the primacy of storytelling: stories, not statistics, are the key to human action. Unless we reflect more deeply on the stories we tell ourselves—and tell others—we'll fall short of the goal, which is action.

Second, there is an overestimation of the importance of knowing—"If only people knew how bad things are . . ." It's taken the field of economics a long time to recover from overestimation of "rational choice" as the primacy of decision-making. Daniel Kahneman won the Nobel Prize in Economics for integrating psychological insights into economics itself.[57] It was right to give him the Prize and to acknowledge so-called behavioral economics as a correction to statistics alone. Maybe economics will catch up with reality.

Advocates around climate change need to learn these lessons; we need to reflect on how thinking both fast and slow shapes what we do and affects how people hear the stories we tell. Storytelling about the apocalypse fails to grasp the importance of hope in shaping how individuals will act. This point is especially important for climate advocacy.

Clarissa Pinkola Estes said it well: we were made for these times:

> My friends, do not lose heart. We were made for these times. I have heard from so many recently who are deeply and properly bewildered. They are concerned about the state of affairs in our world now. Ours is a time of almost daily astonishment and often righteous rage over the latest degradations of what matters most to civilized, visionary people . . .

. . . especially do not lose hope. Most particularly because, the fact is that we were made for these times. Yes. For years, we have been learning, practicing, been in training for and just waiting to meet on this exact plain of engagement . . .

In any dark time, there is a tendency to veer toward fainting over how much is wrong or unmended in the world. Do not focus on that. There is a tendency, too, to fall into being weakened by dwelling on what is outside your reach, by what cannot yet be. Do not focus there . . .

Ours is not the task of fixing the entire world all at once, but of stretching out to mend the part of the world that is within our reach. Any small, calm thing that one soul can do to help another soul, to assist some portion of this poor suffering world, will help immensely. It is not given to us to know which acts or by whom, will cause the critical mass to tip toward an enduring good.[58]

Have we forgotten how Dante, in *The Divine Comedy*, has inscribed above the entrance to hell "All hope abandon ye who enter here"? We need both hope and fear if we are to tell stories that make us all work together to stop climate catastrophe. "Your young shall see visions and your old will dream dreams" (Acts of the Apostles 2:17). We need both young and old together, and, yes, we were made for these times. HERE NOW YOU HOPE!

Notes

1 Charlotte Beradt, *The Third Reich of Dreams: The Nightmares of a Nation, 1933–1939* (Aquarian Press, 1985).
2 From the Metropolitan Museum of Art, N.Y. City at: www.metmuseum.org/art/collection/search/336215
3 On dreams and climate, see Bonnie Bright and Paul Marshall (eds.), *Earth, Climate, Dreams: Dialogues with Depth Psychologists in the Age of the Anthropocene* (Depth Insights, 2019). On apocalypse and dreams see Catherine Keller, *Facing Apocalypse: Climate, Democracy, and Other Last Chances* (Orbis, 2021).
4 For more on dreaming and climate change see Kyla Mandel, "Climate Change Is Changing How We Dream" *Time Magazine* (July 27, 2023) at: https://time.com/6298730/climate-change-dreams/
   See also Samantha Harrington, "Therapist Collects People's Dreams about Climate Change" *Yale Climate Connections* (Nov. 15, 2019) at: https://yaleclimateconnections.org/2019/11/therapist-collects-peoples-dreams-about-climate-change/
5 Rage versus Hope: Why are these two opposed in debates on climate advocacy? It's a false choice: both are needed, as George Monbiot understands:

   "Days of Rage" at: www.monbiot.com/2022/07/19/days-of-rage/

   See also "Outrage and Optimism" from Christiana Figueres for podcasts: www.outrageandoptimism.org/
6 Pema Chödrön, *Start Where You Are: A Guide to Compassionate Living* (Shambhala, 2001).
7 Rebecca Solnit, *Hope in the Dark: Untold Histories, Wild Possibilities*. See also: "Rebecca Solnit on Hope in Dark Times, Resisting the Defeatism of Easy Despair, and What Victory Really Means for Movements of Social Change" at: www.themarginalian.org/2016/03/16/rebecca-solnit-hope-in-the-dark-2/

8 Rebecca Solnit, "We Can't Afford to Be Climate Doomers" *The Guardian* (July 26, 2023) at: www.theguardian.com/commentisfree/2023/jul/26/we-cant-afford-to-be-climate-doomers

9 Ernest Becker, *The Denial of Death* (Free Press, 1997).

10 Velva Lee Heraty, *The Dream Belongs to the Dreamer* (Balboa Press, 2014).

11 Rachel Norment, "Let's Wake Up, Help Save Our World" *Dream Network* 2008;27(3):18.

12 To watch Christiana Figueres talk about her nightmare, see: www.youtube.com/watch?v=KVVW5eGiETI

13 Jason Anthony, "Recognizing the Scale of the Problem" *The Undercurrent* (Apr. 13, 2023) at: https://jasonanthony.substack.com/p/the-undercurrent?utm_source=substack&utm_medium=email

14 Robert Kastenbaum, "Habituation as a Model of Human Aging" *International Journal of Aging and Human Development* 1981;12:3 at: https://doi.org/10.2190/BR5F-H8B7-2B9X-53U8

15 Michael Mann contributed, with other authors, to the award of the 2007 Nobel Peace Prize to the International Panel on Climate Change. Mann was awarded the Hans Oeschger Medal of the European Geosciences Union in 2012 and was awarded the National Conservation Achievement Award for science by the National Wildlife Federation in 2013. More at: www.asc.upenn.edu/people/faculty/michael-e-mann-phd#:~:text=He%20contributed%2C%20with%20other%20IPCC,National%20Wildlife%20Federation%20in%202013.

On Michael Mann, see: D.P. Carrington, "We're Not Doomed Yet': Climate Scientist Michael Mann on Our Last Chance to Save Human Civilization" *The Guardian* (Sept. 30, 2023) at: www.theguardian.com/environment/2023/sep/30/human-civilisation-climate-scientist-prof-michael-mann

See also: www.asc.upenn.edu/people/faculty/michael-e-mann-phd#:~:text=He%20contributed%2C%20with%20other%20IPCC,National%20Wildlife%20Federation%20in%202013. See also, "Not Too Late" from Rebecca Solnit at: www.nottoolateclimate.com/

16 Terry Eagleton, *Hope Without Optimism* (University of Virginia Press, 2015).

17 Hannah Ritchie, "Climate Pessimism Dooms Us to a Terrible Future. Complacent Optimism Is No Better" *Vox* (Mar. 21, 2023) at: www.vox.com/the-highlight/23622511/climate-doomerism-optimism-progress-environmentalism

18 Norm Friesen, "Our Age of Crises Needs Bollnow's Philosophy of Hope" *Psyche* (2023) at: https://psyche.co/ideas/our-age-of-crises-needs-bollnows-philosophy-of-hope?utm_source=Aeon+Newsletter

See also: Ralf Koerrenz and Norm Friesen (eds.), *Existentialism and Education: An Introduction to Otto Friedrich Bollnow* (Palgrave MacMillan, 2017).

19 Casey Gwinn and Chan Hellman, *Hope Rising: How the Science of HOPE Can Change Your Life* (Morgan James, 2018).

20 Thomas Mallon, "What If Nostalgia Isn't What It Used to Be?" *The New Yorker* (Nov. 20, 2023).

21 Mirabai Starr, *God of Love: A Guide to the Heart of Judaism, Christianity and Islam* (Monkish Book Publishing, 2012).

22 Paco Mitchell, "The Coming Storm: Prophetic Dreams and the Climate Crisis" *Depth Insights* (Summer 2016), p. 36 at: www.depthinsights.com/Depth-Insights-scholarly-ezine/the-coming-storm-prophetic-dreams-and-the-climate-crisis-by-paco-mitchell/

23 Ed Simon, "What Viktor Frankl's Logotherapy Can Offer in the Anthropocene" *Aeon* (Feb. 11, 2020) at: https://aeon.co/ideas/what-viktor-frankls-logotherapy-can-offer-in-the-anthropocene

24 Mike Bell, "Comox Valley Climate Change Network" (2019) at: https://dtnetwork.org/wp-content/uploads/2020/08/2019–20-Chronicle-17-Meaning-4.pdf

25 "What to Do with Climate Emotions: If the Goal Is to Insure That the Planet Remains Habitable, What Is the Right Degree of Panic, and How Do You Bear It?" at: www.newyorker.com/news/annals-of-a-warming-planet/what-to-do-with-climate-emotions

26 Climate Dreams (May, 2019).
    Other dreams cited here, unless otherwise noted are from the Climate Dreams Project at: https://climatedreams.com/

27 Climate Dreams (Jan. 1, 2019).

28 Sally Gillespie, *Climate Change and Psyche: Mapping Myths, Dreams and Conversations in the Era of Global Warming* (University of Western Sydney, 2014), p. 110.

29 Joanna Macy, *World as Lover, World as Self: Courage for Global Justice and Planetary Renewal* (Parallax Press, 2021).

30 John Stanley and David Loy, "Occupy the Climate Emergency: Reflections on Climate, Empathy, and Intergenerational Justice" *Tikkun* (Mar. 29, 2012) at: www.tikkun.org/author/a_joint_stanleyj_and_loyd/

31 Robin Wall Kimmerer, *Braiding Sweetgrass: Indigenous Wisdom, Scientific Knowledge and the Teachings of Plants* (Milkweed Editions, 2013). For more on the contribution of Indigenous wisdom see "The Climate Crisis and Aging: Capitalizing on Traditional Knowledge and Innovation" at: www.justsecurity.org/90762/the-climate-crisis-and-aging-capitalizing-on-traditional-knowledge-and-innovation/

32 Jonathan Lear, *Radical Hope: Ethics in the Face of Cultural Devastation* (Harvard University Press, 2008), p. 70.

33 Ibid., p. 117.

34 Anthony Colombo, "The Dream of Climate Change" (Aug. 16, 2017) at: https://dreamspace.io/the-dream-of-climate-change/

35 Climate Dreams (Mar. 3, 2019).

36 To face up to the challenge, see *Activism is Medicine: Health and Relevance for the Human Animal* (2023) by Frank Forencich at: www.amazon.com/dp/0972335811/ref=sr_1_1

37 Just Fran, *Climate Dreams*.

38 Per Espen Stoknes, *What We Think About When We Try Not to Think About Global Warming* (Chealsea Green, 1915).

39 Gillespie, *Climate Change and Psyche*, p. 126.

40 Ward Greene, "Caring for Future Generations" [personal communication], 2022. More about SAGE at: https://wearesage.org/

41 Holly Dabelko-Schoeny et al., "Age-Friendly and Climate Resilient Communities: A Grey–Green Alliance" *The Gerontologist* (Oct. 6, 2023) at: https://academic.oup.com/gerontologist/advance-article-abstract/doi/10.1093/geront/gnad137/7293249

42 Rob de Laet, "Hope from the Amazon" [personal communication], 2021.

43 Stephen Aizenstat, "Tending the Dream Is Tending the World" (2019) at: https://dreamtending.com/wp-content/uploads/2019/02/tendingthedream-new.pdf

44 Roman Krznaric, *The Good Ancestor: A Radical Prescription for Long-Term Thinking* (The Experiment, 2021).

45 Max Zeller, *The Dream: The Vision of the Night*, 1975, cited by Meredith Sabini, "The Field of Dreams" *Dream Time* Winter 2011;28(1):22.

46 Montague Ullman and Nan Zimmerman, *Working With Dreams* (Routledge, 1979); and Jeremy Taylor, *The Wisdom of Your Dreams: Using Dreams to Tap Into Your Unconscious and Transform Your Life* (Penguin/Tarcher, 2009). This entire chapter, and much of this book is an exercise in what Taylor called "projective dreamwork." For more on Taylor's idea of projective dreamwork, see: https://jeremytaylor.com/

47 Edward Edinger, *Ego and Archetype* (Shambhala, 1992), p. 218.

48 Harry R. Moody, *The Five Stages of the Soul: Charting the Spiritual Passages that Shape Our Lives* (Doubleday, 1997). The Wizard of Oz is also a story about how invisible and unrestrained engines of power maintain us in ignorance. The struggle around climate change is one example of the process: Don't miss the man behind the curtain: "These Companies Are Pushing Back on Science Showing Their Pollution" at: www.bloomberg.com/news/features/2023-11-30/pemex-glencore-question-science-behind-satellite-based-methane-tracking

49 Anthony Stevens, *Private Myths: Dreams and Dreaming* (Harvard University Press, 1997).

50 Reza Shah-Kazemi, "Seeing God Everywhere" at: https://thegreenknight.org/booksarticles/writings-2/

51 Rebecca Solnit and Thelma Young-Lutunatabua (eds.), *Not Too Late: Changing the Climate Story from Despair to Possibility* (Haymarket, 2023).

52 Milja Heikkinen, "The Role of Network Participation in Climate Change Mitigation: A City-Level Analysis" *International Journal of Urban Sustainable Development* (Feb. 7, 2022) at: www.tandfonline.com/doi/full/10.1080/19463138.2022.2036163

53 "Give the Next Generation a Reason for Hope" *The Age Buster* at: www.theagebuster.com/blog/2019/4/28/give-the-next-generation-a-reason-for-hope

54 Personal communication.

55 Ted Nordhaus, "Climate Storytelling Has an Apocalypse Problem. The Core Premise—If Only People Knew How Bad Things Are, They Would Change Their Behavior—Seems to Have Left Us Well Short of Our Climate Destination" *Anthropocene Magazine* (Nov. 10, 2022) at: www.anthropocenemagazine.org/2019/06/the-climate-change-apocalypse-problem/

56 Richard A. Clark and R.P. Eddy, *Warnings: Finding Cassandras to Stop Catastrophe* (Ecco, 2017).

57 Daniel Kahnemann, *Thinking Fast and Slow* (Farrar, Straus and Giroux, 2011).

58 Clarissa Pinkola Estes, "We Were Made for These Times" at: www.awakin.org/v2/read/view.php?tid=2195

Index

Note: Page numbers in *italics* indicate a figure on the corresponding page.